UNDERSTANDING DIGITAL RESPONSIBILITIES

Research in Sociotechnical Cyber Security

Series Editors: **Genevieve Liveley**, University of Bristol and **Lizzie Coles-Kemp**, Royal Holloway, University of London

---

This series provides a unique platform for interdisciplinary work that examines the intersection of cyber security technologies and societies. Created in partnership with the national Research Institute for Sociotechnical Cyber Security (RISCS), the series features work produced under the auspices of the Institute as well as specially commissioned books that will be of interest to the cyber security community.

Find out more at:

bristoluniversitypress.co.uk/research-in-sociotechnical-cyber-security

LIZZIE COLES-KEMP AND MARK BURDON

UNDERSTANDING DIGITAL RESPONSIBILITIES

BRISTOL
UNIVERSITY
PRESS

First published in Great Britain in 2026 by

Bristol University Press
University of Bristol
1-9 Old Park Hill
Bristol
BS2 8BB
UK
t: +44 (0)117 374 6645
e: bup-info@bristol.ac.uk

Details of international sales and distribution partners are available at bristoluniversitypress.co.uk

DOI: 10.51952/9781529249798

British Library Cataloguing in Publication Data
A catalogue record for this book is available from the British Library

ISBN 978-1-5292-4978-1 hardcover
ISBN 978-1-5292-4980-4 ePub
ISBN 978-1-5292-4979-8 OA PDF

Cover design: Nicky Boroweic
Front cover image: Adobe Stock/keisuke
Bristol University Press uses environmentally responsible print partners.
Printed and bound in Great Britain by CPI Group (UK) Ltd,
Croydon, CR0 4YY

Bristol University Press' authorised representative
in the European Union is: Easy Access System Europe,
Mustamäe tee 50, 10621 Tallinn, Estonia,
Email: gpsr.requests@easproject.com

FSC
www.fsc.org
MIX
Paper | Supporting responsible forestry
FSC® C013604

Dedication – Lizzie Coles-Kemp
For everyone who has taken the time to talk with us about the harms, frustrations and anxieties experienced when digital responsibilities break down.

Dedication – Mark Burdon
To Sally, Tom and Sophie. Never have responsibilities been so delightful to bear. Love always

Contents

Series Editors' Preface

Genevieve Liveley and Lizzie Coles-Kemp

The aim of this series of short books is to provide the first dedicated platform from which to showcase leading-edge interdisciplinary research focused on *sociotechnical* cyber security. The series features research produced under the auspices of the UK's Research Institute for Sociotechnical Cyber Security (RISCS) as well as specially commissioned works of interest to the national and international security communities. The books published as part of this series recognise that technology-centric approaches to cyber security aren't enough on their own. They introduce the latest research and new ideas for transforming security thinking so that *people* are empowered to be the first and the last line of defence. They address the bigger sociotechnical approaches to cyber security – the cultural, psychological, and behavioural influences that make people act as they do in real world security scenarios.

In this volume, *Understanding Digital Responsibilities*, Lizzie Coles-Kemp and Mark Burdon introduce a novel way of rethinking responsibility in the digital environment. They offer practical approaches to consensus building around issues of responsibility, ownership, and enablement in areas of fragmented digital governance. They provide a richly illustrated roadmap for re-connecting responsibility with the effectiveness of security controls and, in so doing, provide a new way of thinking about more ethical approaches to sociotechnical cyber security.

List of Figures

About the Authors

Lizzie Coles-Kemp is Professor of Information Security at Royal Holloway, University of London. She specialises in the design of accessible and inclusive security approaches and focuses on the design of everyday essential digital services (including housing, health, welfare, education, transport and finance). Since 2008 Lizzie has worked with marginalised and underserved communities investigating how essential services might be better designed to support societal cohesion as well as the security and agency of the individual. Her research practice uses creative engagement methods, and she deploys a participatory research design. Lizzie is Principal Fellow at the UK's Research Institute of Sociotechnical Cyber Security and has a long track record of working in partnership with the UK's National Cyber Security Centre. She is a member of the College of Experts for the UK government departments of Science, Innovation and Technology (DSIT) and of Culture, Media and Sport (DCMS).

Mark Burdon is Professor of Law at Queensland University of Technology. Mark's research areas are privacy law, data protection and the regulation of information security. His research critically examines the complex privacy challenges emerging from the sensorisation of everyday devices and infrastructures. These themes are explored in his monograph, *Digital Data Collection and Information Privacy Law* (Cambridge University Press). He is a Chief Investigator on an Australian Research Council Linkage project aimed at enhancing digital compliance processes within the Australian financial services sector. Mark was a Challenge Fellow on the UK-based Discribe project, which explored the construction of digital responsibilities in environments disrupted by emerging technologies.

Notes on Contributors

Evan Easton-Calabria is Senior Researcher and Fellow at the Fletcher School of Law and Diplomacy and the Feinstein International Center, Tufts University. She also holds positions as a Senior Technical Advisor at the Red Cross Red Crescent Climate and as a Research Associate at the Refugee Studies Centre, University of Oxford. Her research focuses on the structure of humanitarianism and development, including refugee self-reliance, digitalisation, urbanisation and localisation. She previously worked at Royal Holloway, University of London, and the University of Oxford, from where she also holds a Masters and PhD in International Development. She is the author of over 85 publications, including work on trust and digital responsibility, refugees in the digital economy.

Iida Käyhkö is a PhD candidate at the Information Security Group at Royal Holloway, University of London. Her research examines insecurities experienced by political diaspora communities. She has carried out in-depth ethnographic research with the Kurdish community in the UK since 2018, and published on the topic in *The Guardian*, *Race & Class* and *Novara Media*.

Acknowledgements

Funding acknowledgements

With thanks to RISCS and NCSC for creating and promoting a Digital Responsibility Fellowship and for funding Nick Robinson's time to conduct an initial literature review and background research on the Online Safety Act.

With thanks to Toby Scott @ KnowInnovation for his guidance in the first phase of the RISCS digital responsibility fellowship.

With thanks to the DiScriBe programme (ESRC grant number ES/V003666/1) for funding Lizzie Coles-Kemp's and Mark Burdon's time to develop the digital responsibility framework as part of the programme's legal and regulatory research.

Author acknowledgements

Thanks to Evan Easton-Calabria and Iida Käyhkö for their authorship of case studies in Chapters 4 and 5, respectively.

Research assistance acknowledgements

Thanks to Nick Robinson for the initial work on the Online Safety Act case study.

Thanks to Laura Shipp and Nick Robinson for an initial digital responsibility literature review.

Mark Burdon – Thanks to Sasha Ness for excellent research support regarding a literature review of legal responsibility.

Illustrations acknowledgement

With thanks to Chris Day @ Little Creature for the book's illustrations.

PART I

Introducing Digital Responsibilities

ONE

Why Digital Responsibilities?

Why focus on digital responsibilities?

Responsibility is a slippery concept that is difficult to pin down and yet an understanding of responsibilities, how they are allocated and who actions them is fundamental to a functioning society. As societies become increasingly digitalised, understanding how responsibilities are shaped and re-shaped through digital technology is vital if the benefits of digital technologies are to be realised. Digital technology redistributes responsibilities at speed and scale by enabling new and different modes of collaborative working, wealth generation and resource management.

Think of messaging apps such as Signal and WhatsApp and how they enable social and work collaborations over time and space.Think of e-payment platforms and how they enable new forms of marketisation and income generation. These technologies are not used in isolation but are combined in formal and informal configurations to harness new ways of engaging and relating for political, social, business and administrative purposes.

Just as such configurations shape and re-shape how we relate to each other, forming new collaborations and co-operations in the process, so too are the responsibilities that flow through these relationships. If such changes to responsibilities are not realised and established, then the effectiveness of responsible actions is limited. Our work sets out a framework that helps surface responsibilities in digitally mediated spaces and scaffolds conversations about how responsibilities might be realised, established and actioned.

The work originated from a digital responsibility fellowship supported by the UK's Research Institute for Sociotechnical Cyber Security (RISCS). Lizzie Coles-Kemp was the RISCS digital responsibility fellow and theme lead between 2020 and 2023. The fellowship's focus was to set out digital responsibility as a research area, elicit the relevance of digital responsibilities to effective information protection strategies and consult with academics and practitioners on the state of the art in understandings of digital responsibilities. The three case studies included in this book were funded as part of this RISCS fellowship. Lizzie then joined forces with Mark Burdon on the UK's Discribe Hub+ project. Mark was a challenge fellow on this project that was funded by the Economic and Social Research Council. Together they developed the RISCS work on digital responsibility and combined it with Discribe's work on legal and regulatory aspects of digital security by design.

We present that work as our new digital responsibilities framework in Chapter 2. We then explore the framework's capabilities through the three case studies. We conclude by setting out a manifesto for 'doing' digital responsibility in pluralistic, multi-scalar, digital contexts. Issues of power, marketisation, technocratic control and human agency are never far from the surface in our work, as outlined in the book.

The book's rationale

The book outlines a process of enabling collaboration and constructing consensus among the multiple and continuous parties involved in the realisation, establishment and actioning of digital responsibilities. Through a series of case studies, it highlights the dynamic nature of digital responsibility formation that involves multiple parties, with differing requirements and different capabilities. The framework thus recognises that the process of digital responsibility formation is conducted in a genuinely pluralistic environment that features groups and individuals with diverse needs, resources and values.

Our focus on the pluralistic nature of digital responsibilities is novel and important because it allows policy makers to better understand the challenges of governance in environments that are disrupted by digital technologies and infrastructures. The digital environment is pluralistic because it straddles separate sites of responsibility realisation and construction (such as the market, the state and the individual) and across different realms of technology engagement relevant to the establishment of digital responsibilities. In turn, these different sites and realms are governed through multiple plural perspectives – legal, regulatory, infrastructural, computational, etc. – to the extent that not one policy perspective is sufficient to understand the true complexity of digital responsibility formation.

The book is highly relevant to both policy and academic communities. It will allow policy makers to better understand the processes of realising, establishing and actioning responsibility for the digital realm. Governance strategies and policy development need to enable collaboration across separate groups to ensure a genuine consensus about the need and actioning of responsibilities in digital environments. The book outlines to policy makers, and indeed digital technology developers, how to think about policy implementation as an issue of responsibility. It provides a new conceptual understanding of responsibility in the digital realm and provides practical examples of policy fragmentation through three case studies developed by up-and-coming digital researchers from across the social sciences. These works highlight the challenges of actioning responsibility across different digital policies.

The work, by its nature, is multi-disciplinary and will be relevant to different academic disciplines: cybersecurity; policy sciences; law and regulation; digital media and infrastructure. On top of the case studies, the book also brings together several different responsibility concepts (including digital responsibility, legal responsibility, responsible research and innovation) to provide a robust and holistic frame to both better understand and effectively enact digital responsibility from a policy-making perspective.

Different disciplines have differing conceptual understandings and applications of responsibility. As such, it is not possible for the book to cover all dimensions of responsibility, particularly in relation to philosophical perspectives. Issues that are well trodden in the broad philosophical literature, such as moral responsibility or collective responsibility, have therefore not been explicitly covered in the book. They do, of course, feature in the areas of legal and digital responsibility but are done so implicitly, rather than explicitly.

We would like the book to equip policy makers worldwide, and UK policy makers particularly, to have policy discussions about fundamental areas of digital policy including digital identity, service and technology resilience, fraud and scam prevention, and personal data protection.

The book's coverage

The book has three parts: (I) digital responsibility framework overview, (II) three case studies of policy challenges in digitally fragmented environments that surface responsibility issues and (III) a thematic conclusion that brings together the key findings of (I) and (II) to better understand the role of digital responsibilities in policy making.

In Part I, Chapter 2 lays out the main canons of literature in which our work is grounded. Chapter 2 explores academic understandings of responsibility across three frames: digital technology, law and the individual. We build on the extant literature to provide a new way of understanding the complex enmeshing of responsibility formation in digital environments. The process of realising, establishing and actioning responsibility for the digital realm needs to enable collaboration across separate groups to ensure a genuine consensus about the need and actioning of responsibilities.

The chapter concludes by introducing our digital responsibility framework and setting out the high-level and operational process points involved in digital responsibility formation:

- *Realising* entails the realisation and construction of responsibility needs across the different parties involved in digital service delivery and policy development.
- *Establishing* regards the formulation and establishment of digital responsibility requirements, particularly through law or regulation.
- *Actioning* entails how established responsibilities are actioned or non-actioned by parties involved in digital policy implementation.

The digital responsibility framework represents a new way of conceptualising the development and actioning of responsibilities in the digital environment. It outlines the importance of understanding the relations between:

- Realising, establishing and actioning digital responsibilities in a policy context.
- The sites, realms and environments in which digital responsibilities are formed.
- The norms, requirements and values that underpin digital responsibility policy-based formation.
- Enabling collaboration and constructing consensus throughout the processes of digital responsibility formation to better understand and inform policy making.

We contend that aligning these relationships increases the likelihood of successful responsibility adoption which enables the better implementation of policy requirements for digital environments.

Part II of the book provides case studies of digital policy challenges, especially across fragmented regulatory spaces.

The case studies highlight the relationships between realising, establishing and actioning responsibility and the necessary associations between different parties in responsibility generation. These relationships become more important in areas of fragmented regulatory spaces or governance.

The case studies are three individual works that each highlight a digital policy challenge in the following policy areas:

- Development of UK online safety legislation and the fractious process of legislative debate ('Digital Safety') – Mark Burdon and Lizzie Coles-Kemp.
- Lack of much needed digital services for refugees and asylum seekers in the UK ('Digital Migration') – Evan Easton-Calabria.
- Provision of online targeted material to recognise and encourage Kurdish participation in the UK census ('Digital Census') – Iida Käyhkö.

Each case study depicts a snapshot in time. They each show the messiness, complexities and contradictions that challenge the implementation of any framework or policy. The case studies underline the situatedness of responsibility formation and how responsibilities are constructed through relationships and interactions.

As part of the writing process, we handed our draft writing to illustrator Chris Day. While academic writing often flattens or simplifies such complexities in a bid to communicate an argument, we have incorporated illustrations within the case study chapters to draw out some of these complexities. We go on to use illustrations in Chapters 6 and 7 to show how contextual complexities might interact with the framework that we set out. In this way, we hope to ground a concept as abstract as digital responsibility in the day-to-day spaces that stand to benefit from a more holistic approach to realising, establishing and actioning responsibilities in digitally mediated spaces and places.

In Chapter 3, Mark Burdon and Lizzie Coles-Kemp examine the complex and fractious Parliamentary debates involving the development of the Online Safety Act 2023 ('OSA'). The Online Safety Bill 2021 was considered a landmark piece of prospective legislation in its attempt at establishing a new legislative framework for tackling harmful content online. To make the UK a safe digital space, the Bill sought to establish a new 'duty of care' upon online service providers towards their users, obligating them to take positive action to mitigate potentially harmful content. The proposed imposition of this wide-ranging reform was contentious and led to significant debate about the expected responsibilities of different parties. Chapter 3 highlights the digital policy challenge of tackling online harms as a prism to (a) identify how digital responsibility manifests within the different dimensions of private sector digital design, delivery and use and (b) examine the fractious policy debates to identify the responsibility contentions behind the debates. The chapter finds that a responsibility consensus across key parties was not fully realised which meant that legal responsibility was established through the implementation of the OSA which commanded responsible action discharged through technological means rather than it being a consensus driven exercise. The Parliamentary debates thus demonstrate distinct types of responsibility action at play, which were mostly negative, particularly from the private sector parties, in trying to avoid responsibility for online harms and safety.

In Chapter 4, Evan Easton-Calabria examines the roles of various levels of governance, from supra-national (the United Nations) to national (the UK Government) and local (London authorities, civil society organisations) regarding the provision of digital services for UK-based refugees or asylum seekers. The chapter explores several elements of digital responsibility by humanitarian and civil society organisations towards refugees. By the very definition, a refugee is no longer under the responsibility of their home country, but instead has a host state, the UNHCR as a 'surrogate state', or a resettlement country which legally 'takes responsibility' for them. In this light, responsibility in the refugee context is one inherently related to the breakdown of the state-citizen compact or in fact relates to the lack of fulfilment of a (new) state-citizen relationship. The chapter findings highlight refugee-serving organisations in the third sector as important actors filling gaps in state-led refugee assistance, and ones which either explicitly or de facto take responsibility for several aspects of digitally assisting refugees. In doing so, the case study also touches on the intersections between digital responsibility and humanitarian and civic responsibilities. The contexts explored illustrate that just as responsibility occurs in physical contexts, so does it manifest in digitally mediated contexts. In this case, the absorbed actions of civil society organisations provide digital services to refugees with fewer resources to take on such a task, especially during COVID-19 lockdowns.

In Chapter 5, Iida Käyhkö examines digital responsibility practices among London's Kurdish community regarding the 2021 census. The chapter examines the creation of accessible and secure digital services for marginalised communities, particularly those who have a fraught relationship with the state. Käyhkö provides a window into the existing actions of responsibility among London's growing Kurdish diaspora, and what these can tell policy makers about the concerns and challenges encountered in digital interactions with the state. Further, it looks at how digital services are engaged within a particular cultural setting, and why specific cultural environments must be considered in discussions on digital responsibility. The chapter finds that digital responsibility is shaped by specific cultural settings, and that practices of outreach and support may already exist in informal and semi-formal networks, unnoticed or unaided by local councils or central government. Actions taken by community organisers created a sense of trust and excitement around the 2021 census and created secure and accessible environments for community members to complete the census. The shouldering of digital responsibility is therefore not merely a question of providing expertise or assisters to marginalised communities, but rather of engaging with existing community networks from a perspective of civic responsibility.

Part III of the book synthesises the previous chapters to outline future digital policy implications. A 'lessons learned' approach is taken by applying our framework to case study findings to better understand digital responsibility issues.

In Chapter 6, we shed light on the lessons that can be learned when the framework is applied to the case study findings. Five lessons learned are put forward:

- Actions tell a lot about responsibility but not everything. Actions must be understood in the context of realising and establishing responsibility.
- Digital responsibilities are distributed across infrastructures, technologies and governance requirements.
- Responsibility lacunas – an absence of responsibility in certain circumstances – are features rather than bugs of digital responsibility formation and should be seen and supported as spaces for positive digital responsibility formation.
- Using forcing functions, such as legislation or security controls, does not guarantee the outcome of expected digital responsibility actions.
- Resource allocation and revenue generation effect digital responsibility formation at all levels.

Finally, in Chapter 7, we conclude the book by contending that digital responsibilities should be considered in five ways which provides policy makers with a better responsibility-based understanding of the consensus building requirements necessary for implementing policy in fragmented digital environments. These five ways are:

- Understanding the digital is about real relationships.
- Understanding digital impact at scale.
- Understanding that we can't have speed and responsibility.
- Understanding responsibility frameworks differently.
- Understanding that digital responsibility is pluralistic.

Chapter 7 therefore squares the circle and builds on everything that has passed before it. We put forward several different arguments for why digital responsibilities should be understood differently, beginning with the contention that to understand digital responsibility means to truly acknowledge and understand the real relationships involved in digital technology implementation and management.

All the case studies, to a certain extent, demonstrate that individuals can be understood, even by positive intention actors, as something less than an individual – a responsibilised cog of an infrastructural wheel. The Digital Migration and Digital Census case studies demonstrate clearly that even well-intentioned responsibility realisations and establishments require on-the-ground actions that are aimed to meet the needs of the individual. We then contend that it is important to understand the impact and scale of digital technology implementation which helps to explain why individuals become cogs, especially in institutional structures of automation predicated on a hybrid mix of public service delivery and private sector profit.

The impact and scale of digital technology implementation leads to our third understanding. Put simply, responsibility generation, either individualised or through collaborative consensus, cannot be achieved at speed. A conundrum and tension visibly arises in the drive towards automation at scale and the increasing pathways of responsibilisation that cements individuals as cogs of infrastructural implementation. Both the state and private sector technology providers cannot shoehorn responsibility on to individuals, and expect them, at the same time, to be informed and consensual participants

Figure 1.1: The limits of forcing functions

in digital service delivery. We therefore conclude our argument by contending that digital responsibility needs to be understood pluralistically.

When put together, these five understandings shed light on why digital responsibility is different and points to the distinct types of understandings and conversations that are required to better acknowledge the complexity of forming responsibility consensus in digital environments.

What we hope you'll get out of the book

Our book sets out the importance of placing the issue of digital responsibility at the centre of policy making including policies about the digital and policies that rely on the digital for their implementation. Our case studies illustrate the implications of why it is important to consider responsibilities when implementing policies and the harms that can arise as a result. We propose a framework that is intended not to regulate responsibilities but to act as a prompt that scaffolds conversations about digital responsibilities in environments encompassing digital technologies that continually shape and re-shape the interactions upon which our societies are structured.

By using illustrations, we hope to provide a bridge from theory into practice, thereby providing inspiration for how to make the question of digital responsibility central to the design and implementation of public policy in contemporary, digitally mediated settings.

TWO

Responsibility Frames and Digital Responsibilities Framework

The fact we have responsibilities is an axiomatic function of human society. It is unsurprising, then, that responsibility is embedded across different academic literature. Responsibility is considered from perspectives such as political responsibility, climate responsibility, civic responsibility, economic responsibility and organisational responsibility. Moreover, forms of legal and moral responsibility are generally accepted as an ascribable component of everyday life. As a member of a society, we have moral responsibilities as part of our daily functioning that responsible individuals who abide by the multitude of different legal and regulatory obligations are expected to follow. It is not surprising then, from an academic perspective, that responsibility 'admits of a variety of uses' (Fischer and Ravizza, 1998, p 2) and 'all societies are systems of responsibilities' (Barnes, 2000, p 8).

Yet despite its deeply embedded and diverse nature, theorising about responsibility is often grounded in an older model of what responsibility is and how it works. That model, akin to the outcome framework, outlined later in this chapter, understands responsibility as the imposition of external obligation on individuals that gives rise to rational and internal reflection about the consequences of actions. The most influential modern-day exponent of such theorising is the legal philosopher HLA Hart and his understanding of how legal responsibility emanates and permeates other disciplinary outlooks – to the extent that responsibility derived through backward-looking forms of ascription (a core building block of Hart's theories) is considered the norm about how to understand responsibility, whether it be in legal or digital circumstances.

We contend in this book that understanding responsibility solely through processes such as blameworthiness does not fully account for the relational context that responsibilities emerge from. Traditional notions of what individuals are responsible for do not fully capture the plurality of environments and volume of relationships inherent to how responsibilities are formed. Responsibility formulation then is not a simple relationship of societal to individual outcome based on realised needs, established legal consequences and rational individual actions. Instead, responsibility formulation is constituted as a negotiated outcome involving consensus building across multiple parties who recognise the need for collaboration for the benefit of all parties.

Understanding responsibility is consequently not just about individual actions and whether actions are morally or legally sanctionable.

A focus on actions only limits responsibility analyses to the individual and the cognitive abilities necessary to determine consequential outcomes. That framing is no longer sufficient because it does not entail considerations about how responsibilities are realised and established, which we contend are necessary preconditions to thinking about how responsibility is actioned. Digital technologies enable relationships between stakeholders such as between the citizen and government, between employer and employee, between service provider and service user to be negotiated and re-negotiated over time and space, thus attenuating the connections between realising and establishing responsibilities and actioning responsibilities.

Focusing solely on the end point of responsibility misses the importance of situating actions as part of a broader process of responsibility formulation which is linked to how responsibilities are realised and established. These considerations are more so in the digital environment where automation processes continue to disrupt and displace core relationships.

Building on these considerations, our work takes a cross-cutting exploration that examines how digital technology re-shapes our understanding of responsibility across different societal dimensions. We produce a framework that can be used to scaffold conversations about how digital technologies re-shape our understanding of responsibilities and to identify where responsibilities have either not been realised or established or where actions do not meet expectations. We believe such conversations are necessary for more effective awareness of responsibility formulations, especially for digital policy purposes.

The first half of Chapter 2 explores academic understandings of responsibility across three frames: digital technology, law and the individual. The framings are important because they provide the type of holistic coverage necessary to fully understand the complexity of realising, establishing and actioning responsibility. A core theme emerges across each frame entailing the tension between traditional theorising about responsibility with more contemporary notions. Disagreements about whether digital responsibilities are individually focused, or are collectively shared, are resonant in different approaches, such as digital civics or responsible research and innovation (RRI). Similar disagreements arise about whether there is or can ever be a responsibility gap, especially relating to autonomous machines, or whether a human will always need to be responsible. These assertions do, of course, have a deeply legal flavour which emanates in the distinctions between outcome and constitutive frameworks and the different understandings about how legal responsibility is established, either by ascription or by consensus. Both the digital and legal literature also have a position about the role of individuals. We therefore briefly cover individual responsibility to show the different framings of responsibilisation and response-ability, to further outline our understandings of traditional and contemporary approaches.

The second half of the chapter then overviews our digital responsibility framework with its focus on realising, establishing and actioning responsibility. We draw on constitutive approaches to show that responsibility should be considered a collaborative exercise that operates in plural and dynamic environments. Achieving consensus about the formulation and allocation of responsibilities is thus challenging, especially in the digital environment. For us, responsibility is socially constructed and involves how people, groups and societies work and interact with each other to contribute towards the fulfilment of a shared goal. Responsibility formulation then is relational. As Peter Cane rightly notes, responsibilities are as much about who we are responsible to as what we are responsible for. We also provide some brief snapshots of how the book's case studies apply to our framework to foreground the deeper application of case study analysis taken in later chapters.

Responsibility frames

To better understand how digital technologies shape responsibilities, we explore responsibility through different frames relevant to the digital realm. We consider three core responsibility framings involving digital technology, law and individuals. We start with how responsibility is considered in the digital realm, including how responsibility is treated by different frameworks, including RRI. The focus on digital technology shows that responsibility is considered with constrained notions of the relationships involving digital technologies. We then draw on legal scholarship to extend the initial framing of digital realm responsibilities and situate them into the normative realm of obligation generation through law making. Our focus on law shows that there are foundationally separate ways that responsibility is treated in the legal realm. Finally, we consider how digital and legal framings impact upon the responsibilities expected of individuals, as users of digital technologies and as citizens of nation states governed by law. The description and application of the three frames therefore provide different responsibility perspectives and what it entails. We use these perspectives as a way of thinking about what different ideas of responsibility do, and what they do not do, as a basis for our framework formation, highlighted further later in this chapter.

Responsibility and digital technology

The idea of responsibilities related to digital technology is an expanding area of interest. In some literature, the use of responsibility as a guiding frame is contested and responsibility is often used interchangeably with other conceptual concerns, such as accountability, trustworthiness and privacy/security. Nevertheless, the different literature highlights an increased focus related to ethical concerns arising from the ubiquitous use of digital technologies. The focus on responsibility is quite often polar and tends to explore the ethical obligations that should be placed on 'responsible' designers of technology, at one end, and the responsibilities of technology users at the other. The latter are often treated as the referents, arbiters and/or victims of digital (ir)responsibility (Liyanaarachchi et al, 2021; Lobschat et al, 2021). Moreover, the literature tend to consider responsibility and digital technology through a panoply of specific issues rather than a holistic concept that deserves systematic investigation. Such issues include an extensive focus on responsible artificial intelligence (AI) (Lima and Cha, 2020), responsible design and innovation in digital technology (van de Poel and Sand, 2018) and corporate responsibility (Markus and Mentzer, 2014) of digital technology, often referred to as 'corporate digital responsibility' (Dorr, 2021).

Underpinning the focus on corporate responsibility are some core ideas that also permeate through other disciplinary considerations, most notably the tensions emanating from legal versus moral culpability (Lobschat et al, 2021), the assignment of blame and what should constitute blameworthiness (Lima and Cha, 2020). The need to think creatively about how new ethical frameworks could enhance responsible digital technology innovations is also a common feature (Elliott et al, 2021). We note later that the foci on culpability and blameworthiness is a strong element of the legal literature, especially those authors that espouse more traditional notions of responsibility in law. There are limitations to these notions, as we highlight. Some further limitations pertinent to the digital technology literature are also worth noting.

First, considerations of corporate responsibility for digital technology are dominant, which leads to a certain perspective of where responsibility should lie. In other words, the designers of digital technology are solely responsible for their products. We raise this, not to argue for the protection of already over powerful corporations, but rather as a means of highlighting that responsibility considerations are dynamic and plural and thus require the unpacking of complementary concepts, such as obligations, duties and accountabilities which exist across a range of user and provider relationships and networks. For example, state-based digital services are now provided in partnership with corporate partners and other service providers, to the extent that an approach which only focuses on the responsibilities of the latter, especially through the limited focus of accountability based on service level agreements, does not fully capture the complexity of the multi-partner environment. A focus solely on private sector providers does not fully emphasise or encapsulate the civic aspects inherent in the provision of state-based digital services.

Digital civics explores how digital systems and technologies can better support community-centred forms of digital service provision and organisation (Vlachokyriakos et al, 2016). It consequently has a broader focus on responsibility that goes beyond corporate service delivery to consider broader societal values, such as democratic engagement through digital participation (Jensen et al, 2020a). Under a digital civics lens, citizen participation – facilitated by digital technology – enhances the possibility of creating new forms of service model (Le Dantec, 2016a). These new models seek to move service delivery from the purely transactional to consider more deeply the intrinsic relations that are both built through public service delivery (Le Dantec, 2016b; Fan and Fox, 2022) and upon which successful public service delivery is founded. For example, the implantation of public sector values in private sector service provision changes the focus of service users as a citizen-conceptualised activity rather than a consumer-conceptualised activity.

An example of this change in perspective can be seen in work relating to co-creation of services between citizens and service providers (Glick and Crivellaro, 2023). Doing so disrupts the set relationships of 'citizen-as-consumer', 'government-as-data provider' and private sector technology providers as an unquestioned efficiency benefit (Vlachokyriakos et al, 2016, p 1097). The digital civics literature views users primarily as citizens, who should be empowered to act as co-creators and active stakeholder participants in the design and delivery of digital public services (Boehner and DiSalvo, 2016). Digital civics approaches therefore aim to create more participatory platforms, to foster advocacy and move stakeholders towards dialogue and negotiated outcomes generated through consensus (Le Dantec, 2016a). Responsibility, in the digital civics model, thus needs to be considered broadly, as a foundationally relational model of digital service delivery that explicitly encompasses many, and at times conflicting, actor interests.

Digital civics, and its focus on relational services, highlights that responsibility does not exist solely on the part of developers of digital technology. Responsibilities for what happens in the digital realm should also be equally shouldered by service users and those assisting them (Coles-Kemp et al, 2022). The idea of 'shouldering' highlights the considerations of culpability and blameworthiness pertinent to corporate responsibility. Quite often the complexity of digital responsibility environments is often framed as familiar concerns involving security breaches, mishandling of data and other blameworthy trope-like failures. Responsibility in digital technology realms focuses on concerns relating to worst-case errors and catastrophes, thus emphasising the culpability and blameworthy characteristics of understanding responsibility as an issue solely of accountability. Highlighting digital responsibility from a civic viewpoint underscores the possibility of alternative practices and approaches to responsibility construction, including a more positive perspective of responsibility. Another alternative approach that has gained traction is RRI which has been defined as:

> a transparent, interactive process by which societal actors and innovators become mutually responsive to each other with a view to the (ethical) acceptability, sustainability and societal desirability of the innovation process and its marketable products (in order to allow a proper embedding of scientific and technological advances in our society). (Commission: Directorate-General for Research and Innovation and Schomberg, 2011, p 9)

RRI frames responsibility as a social or collective endeavour (Stilgoe et al, 2013). The RRI approach acknowledges, and brings together, the plethora of stakeholders involved in digital technology design. These include researchers, designers, businesses, policy makers and citizens, and RRI seeks to align the potentially disparate values, goals and outcomes of different stakeholders involved in technology innovation. RRI also presents a futures dimension to the idea of collective responsibility for technology innovation based on the collective stewardship of digital technology design (Stilgoe et al, 2013). RRI thus emphasises consensus through networks of value negotiation as an explicit form of collaborative activity.

The collective basis of RRI has heightened the important shift of responsibility focus from the designers of technology towards the actual use of the technology, thus recognising that responsibilities rarely fall on a single person or group. Instead, they are collectively shared. Nevertheless, RRI still has limits as it essentially defines responsibility through stakeholder roles and the expectations that those roles are fulfilled in broader relational contexts (Bellini, 2020) including within the life cycle of technology maintenance. We partly agree with these critiques. The aims of RRI are limited because the approach is focused on responsible actions related to a specific technology or technological innovation, whereas we see responsibility formation in a much broader sense, not simply grounded in a specific technology or technological innovation.

Another popular approach to responsibility and digital technology development is proposed by van de Poel. He highlights how responsibility is used in both a descriptive and normative fashion and unpacks five normative meanings of responsibility (van de Poel and Sand, 2018). A fundamental distinction is drawn between backward-looking and forward-looking forms of responsibility. Backward-looking responsibilities, like the focus on culpability and blameworthiness, typically refers to past actions and the potential attribution of blame or praise to ensure forms of remediation (van de Poel, 2011). Forward-looking responsibilities focus on the proactive steps that may be taken to ensure that certain outcomes, such as the avoidance of harm, occur to others. Again, we discuss similar features of responsibility in our discussion of legal responsibilities later. In a digital context, however, van de Poel is outlining that the increased feasibility for monitoring individual actions means that backward-looking forms of responsibility, based on culpability, are potentially easier to ascribe (van de Poel, 2011, p 47).

Lima and Cha (2020) have extended van de Poel's capability categorisations of responsibility to include notions of authority. Set in the context of AI-assisted bail decision-making, Lima and Cha consider the issues of authority and power with six other varying notions of responsibility that can be ascribed to AI machines. They are: obligation, task, answerability, cause, blame/praise and liability. As we will see, the different categorisations relate strongly to the outcome framework of legal responsibility and its focus on backward-looking ascription. Similarly, Bellini (2020) consolidates the notion of responsibility into three distinct constructs that again relate heavily to the type of backward-looking ascription promoted by van de Poel: duty (being responsible for x); blame (being responsible to y for x) and acting independently (processing responsibilities for oneself, in relation to z).

These different approaches to digital responsibility highlight diverse locus for responsibility analyses. Responsibilities can be considered through responsible design, such as RRI, or through 'responsible user behaviour by outlining the agency, capacity and duties of a user towards others' (Bellini et al, 2020, p 564). Similar themes emerge in the legal literature with the distinction of being responsible for oneself and being responsible to others (Cane, 2002). These differences in approach highlight the shift in responsibility focus away from the designer to the use and the users of technology. Such considerations highlight that understanding responsibility in digital contexts means thinking beyond individual roles to consider the broader relational contexts that govern those roles and the digital technologies that operate within them.

The environments in which digital technology operates have become increasingly complex with the rapid expansion of AI and automation. These advancements have impacted upon debates about responsibility, especially regarding agency and accountability of potentially autonomous systems (Coeckelbergh, 2023; Dignum, 2019). A key concern has been the normative question of responsibility attribution for an autonomous system that causes harm (Coeckelbergh, 2020). Such situations significantly challenge responsibility analyses that seek to ascribe blameworthiness for causal actions.

Responsibility has traditionally been ascribed to the operator or manufacturer of a machine. One way of viewing the dissonance between the scoping of responsibilities and the actioning of responsibilities is the 'responsibility gap', as outlined by Matthias (2004). Responsibility gaps could lead to no human being held responsible for an autonomous system causing harm which could threaten both the moral and legal fabric of modern society. Critics of Matthias' thesis have questioned whether a responsibility gap will emanate and have further argued that even if it does, it will still be able to be reconciled, either legally or morally. Critiques contend that responsibility allocation is a dynamic and flexible process that can 'effectively encompass emerging technological entities' (Tigard, 2021, p 591), including those of a purported autonomous nature.

Questions regarding responsibility and AI have triggered a wave of cross-disciplinary debate in areas of the sciences, philosophy and ethics that has produced further ways of thinking about responsibility. Dignum (2020) alludes to the responsibilities placed on all stakeholders involved in the creation, design, procurement, adoption and use of AI systems. She puts forward a triadic framework for the application of ethical AI-by-design. Other commentators take a different perspective. Lima and Cha (2020) contend that responsible AI fails to accommodate the possibility of holding AI responsible itself. Coeckelbergh (2020, p 2052) also argues that AI technologies do not meet the traditional criteria for full moral agency and thus cannot be held responsible. By contrast, Gunkel suggests a potential framework of 'hybrid responsibility' which distributes responsibility 'across a network of interacting human and machine components' (2020, p 317). Given the multiplicity of positions that have recently developed, it is safe to conclude that the issue of responsibility allocation, especially in the AI context, will be in debate for some time to come.

The recent work into responsible AI highlights responsibility's contested nature and demonstrates the need to consider responsibility holistically and from different disciplinary perspectives. One of our motivations in developing a digital responsibility framework is to identify the elements of responsibility in and through digital technology, not only as an object, but also as an action. Responsibility gaps, should they materialise, can then be identified and actions can be clearly articulated. We now ground this disputed digital technology literature in an older and equally contested literature, that of responsibility and law, which has similar themes to those already discussed regarding the role of normative rule-making and the obligatory expectations placed on individuals and society more broadly. Like its digital forebearer, the ideas of backward-looking blame and forward-looking norm generation are at the centre of key legal differences.

Responsibility and law

Traditionally, law ascribes responsibilities which we are bound to follow or risk sanctioning for non-compliance. Legal responsibility is thus engrained in everything we do. However, for something so axiomatic, understanding responsibility as a component concept of legal application is contested and falls largely into two broad camps: outcome-based responsibility frameworks and constitutively based frameworks. Both frameworks understand legal responsibilities in separate ways: outcome-based frameworks through consequences and constitutively based frameworks through relations. Understanding the differing frameworks are important because, as noted frequently earlier, the literature on responsibility and digital technology consider components of these frameworks tacitly. Before we get that far, we outline the traditional outcome-based understanding and the contemporary constitutively based frameworks of legal responsibility.

The outcome-based framework is highly individualised and focuses on the consequences of outcomes for individuals, and how individuals should think about responsibility in the form of consequences (Hart, 1949; Honore, 1999). Like some of the digital technology authors mentioned earlier, the framework is backward-looking by nature and seeks to historicise responsibility by allocating sanctions for blameworthy actions. The framework is largely retributive and seeks to establish responsibilities with reference to consequential threats (Hart, 2000). Scanlon (2000), for example, outlines two forms of moral and legal consequences that can arise. In the first form, responsibility is attributable in situations where a person can be morally assessed for the blameworthiness of their actions. Attributably, in this regard, can mean imposing moral blame based on previous actions. The second form, substantive responsibility, can lead to the imposition of legal sanctions based on the historical consequences of actions. Both forms apply consequences for past outcomes and the difference lies in the consequence to be implemented, either moral blame or legal sanction.

The historical notion of backward-looking blameworthiness is best captured by Hart's (1949) seminal construction of the four different ways legal responsibility is ascribed. Role-responsibility regards the potential imposition of moral blame or legal sanction for a failure to perform duties relevant to a 'distinctive place or office in a social organization' that are 'attached to provide for the welfare of others, or to advance in some specific way the aims and purposes of the organization' (Hart, 1968, p 212). Hart thus signifies the important relationship between roles and the duties that arise from roles to other role holders and society at large. To behave responsibly, therefore, is to behave as a man (person) 'would who took his (their) duties in this serious way' (Hart, 1968, p 213). Causal-responsibility is attributed to any 'causally efficacious' factor that gives rise to harmful outcome from which liability-responsibility can be imposed through legal sanction (for example, imprisonment for criminal offences or civil penalty fines). Running through all three ways of legal responsibility ascription is capacity-responsibility which refers to the minimum capacity a person must have to be held legally or morally responsible for their actions.

Hart's framework was significantly influential and shaped future discussion about legal responsibility particularly in the context of roles. His work also highlights the importance of individual free will as a key component of the outcome-based framework. How individuals respond to the given circumstances of their lives (luck), the choices they make and whether they have the intellectual capacity to make informed choices are all important determinants for responsibility allocation. Gardner (2003) argued that to assume 'basic responsibility is to assert oneself as a rational being' and to provide justifications and plausible excuses for previous conduct. Responsibility, for Gardner, in its basic sense, is therefore 'exactly what it sounds like: response-ability, an ability to respond' (2003, p 161). It is the ability to use reasons in determining actions, thoughts, choices and wants that defines individuals as rational beings, by being able to give an account of actions in relation to the consequences of outcomes (Gardner, 2003, p 164).

The outcome-based framework deems that individuals of free will and capacity are capable of, and therefore must, under law, take responsibility for their actions and the consequences thereof. Individuals have a responsibility to society to do so as ascribed by the societal application of legal rules. If the ascribed rules are not followed, then individuals must responsibly expect sanction for blameworthy acts. Honore (1999, pp 1–2), in one of the leading works on this framework, argued that 'outcome responsibility means being responsible for the good and harm we bring about by what we do'. As such, 'we live under a system by which a community allocates responsibility according to outcomes, and we are consequently forced to make bets on those outcomes' (Honore, 1999, p 26). Responsibility ascription therefore takes place without the direct collaboration of individuals. Responsibility is ascribed to individuals by being part of society, and if they have capacity, they are responsible for sanction and blame that could arise from the consequences of previous acts.

The contemporary constitutively based model highlights that responsibility is constituted by individuals and groups collectively as part of the fulfilment of social practices. In that sense, it is critical of the core components of outcome-based legal responsibility. For example, Cane (2001) argues that outcome responsibility subsumes the richness of different forms of legal and moral responsibility into a sole assessment of outcomes that is not warranted given the breadth and depth of legal intent. As such, not all forms of legal and moral responsibility are concerned with the ownership of outcomes or the allocation of blame (Cane, 2002). Similarly, Dancohen (1992, p 961) criticises the free will paradigm of traditional models of legal responsibility because it 'treats responsibility as a matter of what we choose to do' rather than 'as a matter of what and who we are.'

Not surprisingly then, the constitutively-based framework is less individualised and focuses on intrinsic social relationships as the basis for understanding legal responsibilities. Rather than focusing on consequences and outcomes, the framework sees responsibility establishment as a process of constant negotiation with other parties involved in social practices (Cane, 2002). The framework can be forward looking in nature and seeks to establish prospective realisations and actions of responsibility through legal norms that guide and promote positive acts. Unlike outcome-based considerations, 'historic responsibility is not an end in itself.' Instead, it 'enforces, reinforces and underwrites prospective responsibility' (Cane, 2002, p 35). Cane's notion of prospective responsibility highlights the intrinsic relationship between backward-looking forms of legal accountability involving historic assessment of outcomes with the forward-looking prospects of law as a positive guide of future individual action. Legal responsibilities thus work historically and prospectively (Cane, 2002, pp 34–5).

The prospective nature of constitutively based frameworks means that responsibility realisation is not solely based on assessments of previous actions. Under these ideas, legal responsibility regards how individuals relate to other individuals and groups involved in social practices involving resources and the core objects of life (Kutz, 2004). The framework considers real-life situations in which responsibilities are formed through negotiations and focuses on the capabilities required to form responsibility consensus. Responsibility is inherently relational and attributions consider social, legal and moral relations between multiple parties involved in specific social contexts (Kutz, 2004, p 587; Cane, 2016). Legal responsibility is always relational, therefore, because at its core it revolves around a three-way relationship between agents, victims and the wider community (Cane, 2002, p 94). It is also functional because different responsibility practices, of which outcome-responsibility is but one, perform 'different, though related, functions' (Cane, 2002, p 103).

The relational and functional components of legal responsibility are inherently pluralistic which means that responsibility arises from negotiation instead of ascription (Dancohen, 1992). As such, rather than being detached from an ascribed responsibility process, the constitutively based framework outlines the active role that individuals play in responsibility establishment through constant negotiations in social practice fulfilment. Responsibility then is based on consensus-driven dialogue and thus the broader responsibility individuals have is also contextually different. It requires an understanding of what responsibilities are needed within specific social practice networks rather than a generalised sense of a responsibility to society. Fundamentally, then, 'responsibility is a social practice and not the neutral registration of independent moral facts' (Kutz, 2004, p 551). Legal responsibility is 'mediated by the existence of widely shared social conventions' (Dancohen, 1992, p 970) which means that 'responsibility must be understood in terms of the ideals of agency and community that they reflect and effect regarding the attribution of responsibility' (Kutz, 2004, p 551).

The importance of understanding the similarities and differences between the two frameworks becomes apparent when considering how the legal responsibility literature links to the digital technology literature. It helps to demonstrate the obligations and expectations placed on individuals and other parties involved in digital technology environments. There are diverging viewpoints on how responsibility is conceptualised in relation to digital systems and technologies. Equally, there are diverging viewpoints in law about the role and allocation of responsibility. There are also some explicit links. For example, Hart's work has been particularly instrumental in contemporary interpretations and taxonomies of responsibility found in non-legal literature to help better understand the different aspects of responsibility in a digital context (Bellini et al, 2020; van de Poel and Sand, 2018). Unpacking responsibility in this way highlights the ways in which digitalisation fragments both our understanding and operationalisation of legal responsibility and offers further nuance about how responsibility is generated and operated in a digital context.

We conclude this brief overview of different literature with how the diverse notions of responsibility that arise from digital and legal contexts impact upon individual users of digital services and technology.

Responsibility and individuals

Let's recap before moving on to our final responsibility consideration. We have considered responsibilities that arise in relation to digital technologies. Different responsibility approaches have recently developed which seek to establish responsibilities that are mediated through technology and technological infrastructure, with a particular focus on design. Responsibility construction, in this sense, is inherently relational and relationships are constructed through digital means. Attempts to understand responsibility intertwine with similar concepts found in the legal discipline, namely, ascription of obligation especially regarding backward-looking considerations of blameworthiness and the establishment of forward-looking norm setting, where responsibilities are negotiated through social practices. Responsibilities in the digital realm focus on relations governed through technology and responsibilities in the legal realm govern through rules and norms that ascribe and guide.

The individual is the centre point of both considerations, whether they be constructed as 'user', 'consumer', 'citizen', 'litigant' or 'accused'. As we argue later in the book, the infrastructural context of the digital realm is such that responsibility actioning is a shared activity that encompasses multiple parties. Consequently, it is no longer sufficient to think about responsibility from a purely individual sense, and the implication that individuals bear the primary focus of responsibility. Instead, it is more fitting to examine questions of responsibility collectively because they impact everyone (Ognjenovic, 2010, p 2).

The individual, as a construct and an entity, remains an intrinsic component of understanding responsibility. However, an overt focus on responsibility individualism, and the forms that it takes, does not provide a sufficient lens to consider the full richness of responsibilities in the digital realm. With that starting point, we go back to consider the roots of responsibility and explore what expectations there are on individuals to make moral choices for themselves and the societies surrounding them.

Like what we have already seen, there are contentions about the varieties of moral responsibility. It is clear, from the previous discussion, that even though responsibility in the digital and legal context is different, they both nevertheless draw on and have been foregrounded by notions of moral responsibility. For example, to use Cane's differentiation of responsible for and responsible to, the historical focus of moral responsibility, with its emphasis on 'moral agents', has concentrated on the former. In other words, when should individuals be responsible for certain actions, especially involving culpability and blameworthy assessments? The basis of moral responsibility, much like the backward-looking approaches of the digital context and the outcome-based approach in law, seeks to ascribe responsibility for historic actions that are determined by the cognitive capabilities of a given individual. Responsibility for individual actions is thus a capacity assessment in which individuals can be held to moral account for their actions. Accountability is a concept that resonates across moral and legal plains with the difference between the two hinging on law's structured institutional capacities to render individual account through enforceable legal rules. Moral sanctions, while carrying with them the prospect of similar societal turpitudes, do not have the same institutional characteristics or mechanisms of retribution and punishment (Cane, 2002).

Our fleeting interest in moral responsibility is used to show a nature of responsibility theory that pervades both the legal and digital which privileges personal account. Personal account regards the question of '*what should I be responsible for?*' rather than '*who should I be responsible to?*' It refers to how individuals are ascribed responsibilities for what they should be accountable for. We argue that responsibility in the digital realm is more complicated and there is a distinction between being held accountable and bearing responsibilities. We contend the latter involves negotiated possibilities and consensus formulation that is beyond the traditional processes of being held to account, either legally or morally. Both give rise to complex considerations and even more complex conversations.

For now, we show that being held to account as part of responsibility ascription highlights the degree to which personal responsibility is built within the relational contexts of other forms of responsibility. The degrees that other forms of responsibility are built on, and deem individuals responsible by being accountable, highlight the singular base from which responsibility is ascribed and why it is ascribed.

A question persists at the heart of discussions about ascription, accountability and responsibility that regards who has the power to allocate responsibility and who benefits from the ascription of responsibility to hold account. These degrees and distinctions are explored through two different contemporary responsibility approaches: responsibilisation and response-ability.

Responsibilisation refers to the tectonic movements relating to the transition from the post-war liberal order to the neoliberal state. A key movement is the shift from government to governance (Rose, 2000) with its increased focus on decentralised authority resulting from privatisation of public services (Shamir, 2008). Decentralisation also seeps down to individuals with changing expectations about the capabilities of individuals to determine their own best interests and pursue them through rationalised goals and actions (Peeters, 2013). A core aim of the neoliberal state is to provide greater freedoms of choice for individuals as a core justification for privatised services. The monolithic delivery of essential government services is replaced by competitive private sector innovations, which means that citizens will be able to better match service delivery outcomes to their own needs. However, as a number of authors have highlighted, greater freedoms of choice paradoxically lead to new forms of control by making individuals responsible for choice management (Bogard, 2005; Castells, 2009; Davies, 2018; Deleuze, 1992). The benefit of increased freedom of choice thus comes with the overburdening of self-actualisation and self-management (Trnka and Trundle, 2014).

We see then a different form of personal responsibility from that of the moral actor to the governed individual. The latter is no long ascribed responsibility by being held to account for blameworthy actions. Instead, they are over divested with responsibilities for choice that require insuperable capabilities to fully realise those choices, including the incalculable risks that could arise from choice selection. A final point is also worth noting regarding the shift from singular service delivery through the state to multiple forms of service delivery via privatisation. Responsibilisation strategies that divest state responsibilities means that individuals become the target group for responsibility allocation, and if not, then new collectivities need to emerge (Trnka and Trundle, 2014). The dynamic nature of shifting responsibilities and new organisational emergences means that relational contexts increase in importance. However, responsibilisation has a bald view of this relational context as it stratifies responsibility to designated parties, most notably the individual. Response-ability approaches, on the other hand, have a different perspective about the complexity and importance of relationships, understood broadly and collectively.

At a surface level, it is possible to draw parallels between the individualised focus on moral responsibility and the collectivist orientations of individual outlooks considered by response-ability scholars, as they both, to a degree, regard the capabilities of individuals. Any surface level similarities are quickly countered because response-ability notions reject 'the metaphysics of individualism' and instead focus 'collective knowing, being and doing' (Bozalek and Zemblylas, 2023). Understanding the collective capabilities of individual entities requires responding with care because each component of life, including humans, cohabits and is intimate with each other. Being response-able regards responsive awareness to environs. Unlike the culpability assessments of backward-looking forms of ascription or the drowning of cognitive choice under responsibilisation, response-ability approaches focus on ontological entanglements with 'the other' rather than focusing on individual conscious intent (Bozalek and Zemblylas, 2023). Emergences of responsibleness are generated through the multi-directional relationships of social practices which requires a cultivation of attentiveness to noticing 'the specificity of details in everyday practices' (Bozalek and Zemblylas, 2023, p 65). Being responsible entails a range of different adjectival actions beyond rationality and cognitive capacity. They include: care, attention, openness, acceptance and responsiveness. Response-ability approaches then 'emphasises how crucial cultivating the ability to respond and to allow the other to respond is for collective flourishing' (Bozalek and Zemblylas, 2023, p 78).

In conclusion, the ideas presented in our brief tour of responsibility theories highlight diverse ways of thinking about responsibility. All aspects of responsibility analysis have at their core a construct of the individual, but as we have seen, that construct can be responsible in different ways and for different purposes. We have shown the transition of responsibility considerations based on individuals from self-management to self-capable to response-able which intersects with digital and legal literature. We now build on these ideas to present a new digital responsibilities framework that outlines the complexity of responsibility formation in digital realms.

Digital responsibilities framework

Building on the previous discussion, we now present our framework for better understanding the process of responsibility formation in the digital realm. The framework acknowledges that digital responsibilities are different to the non-digital world. The mediation impact of digital technology is such that it disrupts the relational contexts in which responsibilities emanate. In that regard, our framework is based on the forward-looking nature of responsibilities and draws more inspiration from the collective role of responsibility generation than backward-looking ascription. As you will see, for us, constructing digital responsibilities is a consensus building activity that involves the alignments of different actors, values and activities to provide successful digital implementation strategies. The idea of alignments and responsibilities is not new, as outlined in our coverage of RRI. However, rather than focus on responsibility as a component of specific technology design and use, we seek to scaffold conversations of responsibility in contexts where technologies are combined and re-combined by different and emerging parties to create many and varied digital capabilities.

Our framework is broader because it considers the capacities, capabilities and spaces where digital responsibilities form and as a result also captures the infrastructural consequences of digital implementation. The infrastructure, and the technology it enlivens, both connects and distances individuals concurrently (Acker, 2025). Understanding connection and distance therefore requires an understanding of governing spaces and the activities that take place within those spaces to better understand the relational context of responsibility formation. As alluded to already, our framework is distinctly relational, and we agree with the authors noted earlier that to understand responsibility formation is to understand relationships. How responsibilities are realised and actioned, principally through digital infrastructures and technologies, gives insight into how different actors – users, designers, regulators – relate to each other and could form consensus around responsibilities.

Our framework outlines a process of enabling collaboration and constructing consensus among the multiple and continuous parties involved in the realisation, establishment and actioning of digital responsibilities. The framework represents the dynamic nature of digital responsibility formation that involves multiple parties, with differing requirements and different capabilities. The framework thus recognises that the process of digital responsibility formation is conducted in a genuinely pluralistic environment that features groups and individuals with diverse needs, resources and values.

The digital environment is pluralistic because it straddles separate sites of responsibility realisation (for example, the market, the state and the individual) and across different realms of technology use relevant to the establishment of digital responsibilities. In turn, these different sites and realms are governed through multiple plural perspectives – legal, regulatory, infrastructural, computational, etc. – to the extent that not one perspective is sufficient to understand the true complexity of digital responsibility formation. A pluralistic kaleidoscope is required to view sufficiently the complex enmeshing of distinct perspective requirements, values and norms that are an intrinsic part of the parties/individuals involved in responsibility formation.

The process of realising, establishing and actioning responsibility for the digital realm thus needs to enable collaboration across separate groups to ensure a genuine consensus about the need and actioning of responsibilities. Figure 2.1 outlines the high-level operation of the framework.

The framework identifies three process points involved in digital responsibility formation:

- *Realising* entails the realisation and construction of responsibility needs across the different parties involved.
- *Establishing* regards the formulation and establishment of digital responsibility requirements.
- *Actioning* entails how established responsibilities are actioned or non-actioned.

Each process represents different requirements of responsibility formation and each one is governed by the aims of enabling collaboration and constructing consensus. As such, while each process is distinct from each other, they are nevertheless built upon values of collaboration and consensus construction that recognise the pluralistic nature of the digital environment. The framework thus seeks to sufficiently capture plural dynamism rather than present a linear flow process that results in a simple 'input to output' decision-making mechanism. As already noted, backward-looking forms of ascription simply do not match the conceptual complexity that we are calling for.

In our framework, some responsibility realisations may never come to full fruition, and some that are, may not become established or actioned. The framework therefore seeks to recognise the essential contests involved in digital responsibility

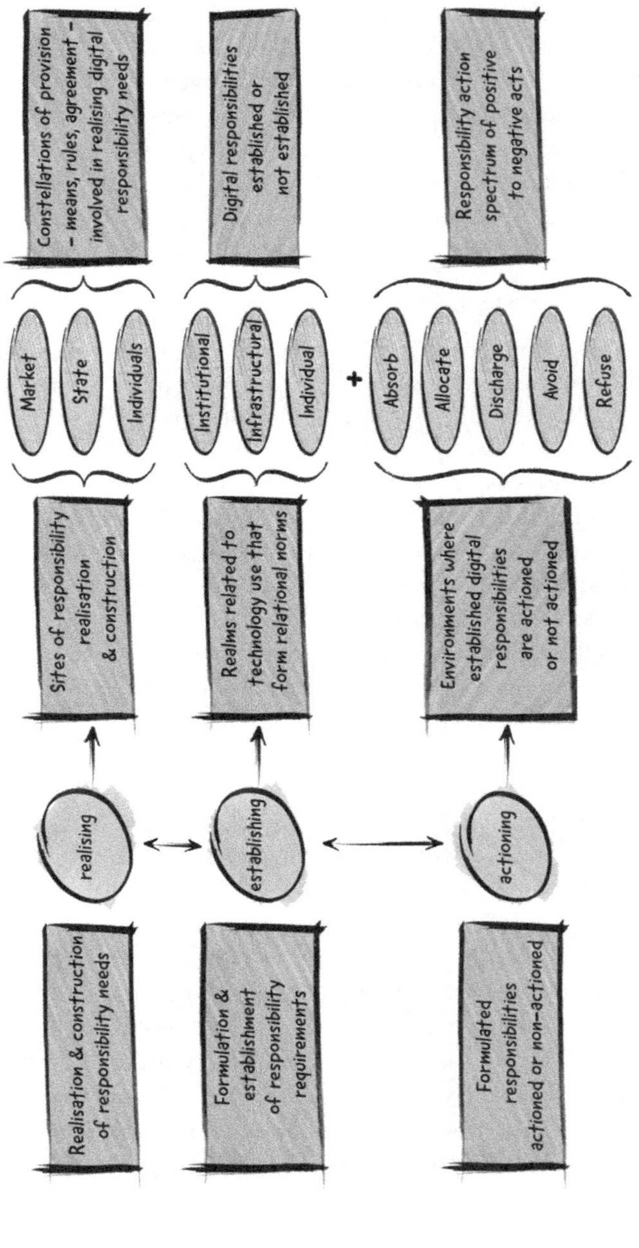

Figure 2.1: Our digital responsibility framework

formation that can impact upon each process, in diverse ways and at contrasting times. Again, we do this to highlight the multiplicities entailed in digital responsibilities and to acknowledge the genuine need for consensus building processes as part of the overall understanding of responsibility formation.

Each process component – realising, establishing and actioning – is now briefly outlined.

Realising responsibilities

As we set out at the start of Chapter 1, our work is grounded in the digital responsibility research programme sponsored by the UK's Research Institute for Sociotechnical Cyber Security (RISCS). The programme's initial review of the UK online harms debate surfaced two important dimensions of the digital responsibility debate: sites and realms. The term *site* refers to a constellation of institutions that share a similar set of digital responsibility considerations and concerns. Our initial analysis showed that responsibilities take the form of duties and obligations and are primarily implemented in one or more sites of responsibility. Our analysis further showed that in the digital setting, no one site of responsibility can ascribe and discharge digital responsibilities without the collaboration of the other sites of responsibility. At the same time, our review of the debates indicated that responsibility is discharged in two distinct areas of activity: (i) use of digital technology to create and curate digital content and (ii) design and development of technology. These areas of activity we term *realms* of digital responsibility. In each of these realms, we observed that security technologies are included within the digital design to ensure that the user of the technology discharges digital responsibilities.

The realisation of digital responsibility is conducted across different sites. Drawing inspiration from Dunn Cavelty and Egloff's (2019) analytical framework used to identify the interplay between main areas of tension in cybersecurity policy, the initial review of the online harms debate set out the three key sites that shape responsibilities: the *market*, the *state* and the *individual*. These sites and the tensions between them have shaped the design of our digital responsibility framework.

Each site provides varied reasons, wants and capabilities involved in realising the need for new or updated digital responsibilities. Each site also provides different foci necessary to the responsibility realisation and construction process:

- *The market* represents the infrastructural and computational requirements of digital responsibility engagement which is now significantly in private and commercial hands, both in terms of operation and future development. It provides the *means* for realising the need for digital responsibility formation.
- *The state* represents the governance requirements of infrastructural existence and operation. The state, its body of regulators and law makers, establishes the governance framework in which digital responsibilities are realised and established. It provides the *rules and guidance* for realising the formation of digital responsibilities.
- *The individual* represents the consensus requirement necessary for realising and constructing digital responsibilities. The incorporation of the representative individual as part of the responsibility realisation process ensures that the means of the market and the rules of the state are designed for the needs of populace consensus. The individual provides the *agreement* for realising the construction of digital responsibilities.

The framework identifies that outcomes of realisation and construction will always be different and it is therefore representative of the differing plural interests at play across all sites. These different outcomes are called the '*constellations of provision*'. In some realisations, and constructions of those realisations, the interests of one site (for example, the state) may outweigh the others (for example, the market) and vice versa. Responsibility realisation thus involves consensus generation across all sites with the negotiated capacity to recognise and acquiesce to different interests for distinct reasons. However, there must be agreement to assist the effective establishment and actioning of digital responsibilities.

Establishing responsibilities

Once digital responsibilities are realised, we contend that the next process regards establishment of responsibility. It is here where digital responsibilities are formulated and established in realms related to technological and infrastructural use. Relational norms shape different dimensions of digital responsibility activation by different parties, in diverse ways. It is in these realms where institutional, infrastructural and individual responsibilities become established or not established. As such, even if the realisation of responsibility is identified and constructed, it does not automatically mean that it will be established. It will only be established through the processes of collaboration enablement and consensus construction identified in our framework.

The three case studies that make up the middle body of the book all demonstrate the importance of the establishment process but in separate ways. Chapter 3's exploration of the debates that preceded the Online Safety Act shows that establishing responsibilities forcefully through ascription of legal rules resulted in a defined legislative outcome which did not receive the full support of all parties, most notably online service providers. Forceful establishment thus has its limitations, even when it is the ultimate force of state-based legislative development. Chapter 4's coverage of asylum seeker digital needs shows the opposite. An international legal obligation was established upon state governments to provide digital access with broad consensus about its need and benefit for vulnerable persons seeking to live in a distant country. That consensus changed over time, however, to the extent that even though a responsibility was established and recognised by all, it was still not actioned. Finally, Chapter 5's coverage of how Kurdish communities in London responded to the census, again shows an established responsibility through law that met differing levels of support from different actors. All three case studies highlight the significant role of legal and regulatory rule formation as part of establishing responsibility.

It is therefore important to understand the role that responsibility plays in legal application because it is crucial in how law establishes individual and collective understanding of responsibility actions through different frameworks, namely, the outcome-based and constitutively based frameworks. Each framework has processes that signify different responsibility outlooks for responsibilities that are established, understood and actioned involving law. Each framework also aims to produce different identity type individuals that will lead to several types of responsibility perceptions and actions. The outcome-based framework seeks to produce the rational individual, a person of free will who is able to make legal and moral choices based on ascribed obligations. Hence the framework's focus is individualised and involves elements that regard individual and internal considerations regarding consequences of outcomes.

We consequently believe that identity formation is at the heart of establishing legal responsibilities. Understanding the role of legal responsibilities better identifies the role of law in the establishment of individual and collective identities. Establishing responsibilities thus has a bifocal effect akin to Cane's differentiation of responsible for and responsible to. Legal rules establish responsibilities through ascriptive means. Legislation provides the framing for which blameworthiness can be identified, articulated and actioned. That track, as noted earlier, is well worn. Equally, however, legal rules also provide the frame for establishing governable identities; for example, a citizen or corporation that is compliant or non-compliant with the law. To do so requires individuals to understand who they are and how they interpret their role individually and in collectives. Individuals thus subsume external legal obligation internally to make rational choices about consequence. Identity formation is established internally within individuals (for example, consciousness) and externally in the context of broader societal expectations and governance. There is a relationship, then, between responsibilities, the expected fulfilment of responsibilities and the understanding of individual and collective identity formation in the realisation, establishment and actioning of responsibilities.

However, as outlined earlier, each framework has a different mode about how legal responsibility is established, especially from the perspective of internal identity construction. Under the outcome-based framework, individuals of free will and capacity take responsibility for their actions. Individuals have a responsibility to society to do so as ascribed by legal rules. That is why the outcome-based framework seeks to establish the rational individual – a person of free will who is able to follow legal responsibilities based on an understanding of established consequences, including the sanctions and blame that could flow. On the other hand, the constitutively based model highlights that responsibility is assembled by individuals and groups collectively as part of the fulfilment of social practices. The framework seeks to produce the intersubjective individual (Dancohen, 1992) – a person who accepts responsibility by understanding their relation to others during the fulfilment of social practices, and their role in the ongoing consensus-focused negotiations that take place across networks. The framework's focus is on understanding responsibility formation as an ostensibly external process involving active interaction with social practices that generates the need for negotiated responsibility outputs.

A foundational difference thus exists between both frameworks. In the outcome-based framework, responsibility is established through law and the locus of analysis shifts to whether the ascribed legal rule has been complied with. Establishment, as seen through some of the case studies, can be forced but may not automatically lead to successful implementation. In the constitutively based framework, establishment is constructed with the individual being an active participant in their understanding of the responsibilities that emerge from active participation involving social practices, including law.

Both processes have similar characteristics that relate to internally generated understandings of responsibility and externalised responsibility actions. However, the ordering of internal and external processes is different across both frameworks. The outcome's framework is unidirectional and that sees responsibility formation by individuals through law as a linear and set process of external and internal components. External law ascribes legal obligations and individuals subject to those obligations internalise them rationally. A consequential assessment is then made internally to ascribe with the obligations and to take responsibility for actions that then flow. Responsibility establishment is thus formed externally and understood internally.

The constitutive framework's consideration of external and internal components is multi-directional and dynamic. An established legal obligation can still exist but instead of directly ascribing individual responsibility through an internal process, an individual's understanding of responsibility is determined through their interaction with others in given social practices. Rather than a clearly delineated process of external to internal responsibility generation, the process of constitutively understanding responsibility combines both external and internal components together, at the same time. Responsibilities are therefore understood through active interaction with others, and understandings of external environments, rather than a purely internal process of rational free will. In that sense, establishing responsibilities is a relational act based on how individuals relate to each other across networks and infrastructures.

Identifying how responsibilities are established is therefore important because doing so shows that, like the process of realisation, responsibilities can be established in different ways, even when the process of responsibility establishment through law looks facially similar, as the three case studies identify. As noted earlier, each case study has a different type of established legal obligation that provides a range of responsibilities for different parties. However, how those responsibilities were understood were markedly different across the parties. Understanding responsibility establishment thus bridges responsibility realisations and foregrounds responsibility actions. We contend that understanding establishment as the realisation of responsibility consensus helps to better identify and understand the concomitant actioning of responsibility by individual actors in collaborative settings.

Actioning responsibilities

Unsurprisingly, the final process regards digital responsibilities that have been realised and established and now need to be actioned. The process represents different approaches to responsibility enactment which gives rise to different types of responsibility act which are conducted in the physical and virtual environments relevant to the actioning of established digital responsibilities.

We contend that responsibility acts work across a spectrum of positive and negative actions. Accordingly, our framework recognises that parties and individuals undertake some responsibility acts positively, some are begrudgingly undertaken and others are avoided or refused. As such, even if a digital responsibility has been realised and established, its enactment cannot be taken for granted as automatically flowing from realisation and establishment. There is still the continual requirement for collaboration and consensus that is necessary to enable positive responsibility actions. Equally, because digital responsibility formation is conducted in a pluralistic environment, there also must be collaboration and consensus across sites, realms and norms, and aided through legal, regulatory infrastructural and computational requirements.

Again, the case studies demonstrate the dynamic nature of these considerations. In each case study, there are established responsibilities through different forms of legal requirement, namely national legislation, international treaty obligations and domestic regulatory requirement. Nevertheless, how parties involved in each case study actioned their responsibilities differed markedly. Some parties, most notably the Kurdish civil sector groups in Chapter 5, were willing to take on extra responsibilities to ensure that potentially helpful legal obligations arising from the census's implementation were better received and actioned across the Kurdish community in London. The same could also be said for civil society groups working with refugees and asylum seeker communities in Chapter 4. Equally, however, the opposite forms of action occurred, involving private sector online service providers in Chapter 3 and a host of national and local government organisations in Chapters 4 and 5. The case studies then demonstrate that the actioning of responsibility should never be taken for granted as either realised or established by all parties. One process can influence the other and that also regards the actioning of responsibility, especially in digital implementation strategies, when existing relationships can be displaced or disrupted through the advent of innovative technologies.

Our framework considers five types of non-exhaustive act that are identified on the positive to negative spectrum through the case studies. The individual or party responsibility actions are as follows:

1. *Absorb* – in this situation, parties willingly take on responsibilities designated to them and potentially designated to others who may not have the will or capabilities to act sufficiently upon their established responsibilities. Absorption is the highest positive act as it seeks to fulfil

the agreed consensus of the environment at potentially the expense of the individual party's own benefit. A key act of absorption arising from the case studies is the actions of the various Kurdish civil society groups who absorbed extra education and training responsibilities to ensure new census requirements were better met.

2. *Allocate* – in this situation, parties take on responsibilities designated to them, and to others, to identify and share those responsibilities across environments. The act of allocation recognises that a party is willing to take on responsibility, but rather than absorb it, and enact the responsibility itself, it designates to, or negotiates with, another party for actioning. Allocation is still a significantly positive act because to allocate is based on (a) ensuring the furthering of a responsibility act and (b) that furthering is based on collaboration and consensus. Again, the co-ordination of community response, and the sharing of resources in relation to positive responses, was a key similarity in the Kurdish and civil society actions outlined in Chapters 4 and 5.
3. *Discharge* – refers to the baseline act of fulfilling or discharging a responsibility. The act of discharge sits in the middle of the spectrum because it can represent both positive and negative actions. A discharge can be positive because, as a base requirement, there is agreement to take on responsibility. It can also be a negative act in situations where some parties could clearly do more to undertake a responsibility; for example, it has the capabilities to absorb or to allocate, but it chooses not to and thus merely discharges, rather than extends, its minimum responsibility. A discharge is thus contextual and representative of the pluralistic environments of digital responsibility formation. An example of discharging behaviours arising from the case studies involves minimum level implementation strategies for asylum seeker digital services provided by local authorities. The same could be said of the local authority actions relating to the Kurdish community and the census.

4. *Avoid* – refers to the negative act of avoiding a responsibility that has been realised and established as relevant to a party. Avoidance is a negative act because it is likely to result in a non-action which has been identified by parties in the responsibility environment as important for the party to undertake. As such, acts of avoidance start to impinge upon the collaboration and consensus requirements inherent to digital responsibility formation. The nature of avoidance is such that there is not a direct refusal to collaborate and negotiate with other parties. Rather, there is a singular focus to act in the best interest of the individual party over the needs and values of environmental collaboration and consensus. As Chapter 3 highlights, many of the contentions arising from the OSA's fractious debates regarded online service provider attempts to avoid responsibility for the broader protection of digital environments.
5. *Refuse* – a definitively negative act of refusing to take on a responsibility that has been realised, established and actioned by other parties flowing from collaboration and consensus. Refusal is an act that conflicts with the expectations of parties involved in environmental responsibility formation. It results in a negative effect across the environment, the sites and the realms of digital responsibility formation, because it directly goes against established norms, requirements and consensus. It also results in a defiantly negative outcome of dis-establishing responsibilities that have been willingly accepted by others in the responsibility environment. The case studies show that the impact of avoidance and refusal acts means that realised and established responsibilities eventually emanate on individuals through responsibilisation strategies. As such, even though none of the case studies specifically involved a refusal action, the extent that different parties tried to avoid responsibility had the same sort of impact upon individual citizens, namely, to try to make them responsible for someone else's actions.

Conclusion

Our digital responsibility framework outlined previously represents a new way of conceptualising the development and actioning of responsibilities in the digital realm. It outlines the importance of understanding the relations between:

- Realising, establishing and actioning digital responsibilities.
- The sites, realms and environments in which digital responsibilities are formed.
- The norms, requirements and values that underpin digital responsibility formation.
- Enabling collaboration and constructing consensus throughout the processes of digital responsibility formation.

We contend that it is the alignment of these responsibilities that increases the likelihood of successful responsibility adoption. At the heart of our framework is the notion that responsibility is an essentially collaborative exercise that operates in plural and dynamic environments. In that sense, it draws strongly from the forward-looking basis of law's constitutively based framework and sees the role of responsibility as ultimately a means to foster consensus. Being responsible then is not simply about being responsible for something, whether that arises from a moral or legal standpoint based on ascriptive means. Responsibilities, in that sense, do not simply arise through roles, causal effects or liabilities. Instead, the formation of responsibilities should be shared beginning with the acknowledgement that collaborations are essential to the better implementation of legal, regulatory, infrastructural and computational requirements for digital environments.

We now move to the book's case studies which outline three different implementation strategies relating to digital capabilities and the uncertain flow of responsibilities that then ensues from implementation. As noted in Chapter 1, each chapter has illustrations to highlight key points.

PART II

Digital Policy Challenges

THREE

Digital Safety Debates and Regulatory Environments

Introduction

The Online Safety Act 2023 (OSA) is a landmark piece of legislation that attempts to establish a new regulatory framework to make the UK 'one of the safest places to be on the internet' (Department for Digital, Culture, Media & Sport and Home Office, 2019). The OSA establishes a new 'duty of care' for digital service providers towards their users, obligating them to act against harmful content published on their sites/platforms and to provide appropriate protections for users, based on vulnerabilities. It is unsurprising that the OSA's implementation was deemed controversial, especially by technology companies, given its wide-ranging regulatory requirements for digital service providers, such as social media platforms and other major technology companies. The significant regulatory change brought about by the OSA is captured by Nash and Felton who rightly contend that the Act has 'tried to embrace an ambitious new model of regulation that could rebalance corporate conduct amongst the biggest technology platforms in favour of the public interest' (2024, p 13).

Given the change required, it is unsurprising that tensions and contentions surfaced during the OSA's pre-legislative development phases arising from the publication of an initial White Paper (Department for Digital, Culture, Media & Sport and Home Office, 2019) that formed the basis for both a draft

Online Harms Bill in 2021 and then a final Online Safety Bill in 2022. Many of these tensions involved the issue of who will take responsibility for implementing the required regulatory measures, including efforts by key industry stakeholders to shape future legislative application in line with perceived interests and business models. The primary areas of contention centred around issues of content moderation, platform design (and algorithmic accountability) and the individual/collective harms that are created by these platforms and their content, as well as their effects upon society at large.

Many of these debates were captured in the more than twenty Joint Committee hearings where several key stakeholders (politicians, regulators, rights campaigners, researchers and technology companies) discussed these issues in-depth and allowed for the scope and expected impact of the Bill to be tested through expert witness testimonies. From these exchanges, the debates around the Online Safety Bill, and online harms in general, are interwoven with discussions around duties, obligations and the accountabilities of certain individuals/stakeholders.

This chapter examines the fractious debates entailed in the Bill's passing. Core tensions are identified through a thematic analysis of the Bill's hearings published in Hansard. Two specific areas of contentions are investigated: (1) age verification and (2) scam advertising. Examination of key testimony in these two contentious areas highlights different attempts by the state and industry to allocate and avoid responsibility. The analysis points to how digital responsibility is established and realised at state-level. It identifies how digital responsibility manifests within the different dimensions of digital product design, delivery and use by the private sector, including different responsibility contentions that underpin the debates. Doing so places a spotlight on the roles that the core technologies of authentication and access control play in forcing the actioning of digital responsibilities by both the end users of digital technologies and by technology providers.

The chapter finds that a responsibility consensus across key parties was not fully realised which meant that legal responsibility was established through the implementation of the OSA. Legislation was used to essentially command that responsibility was discharged through technological means. The Parliamentary debates thus demonstrate diverse types of responsibility action at play, which were mostly negative, particularly from the private sector parties, in trying to avoid responsibility for online harms and safety.

Background

The broader policy intention of the Online Safety Bill was to create an international benchmark and gold standard for online regulation of harmful content, at a time when it was becoming clear that harms were increasing and self-regulation was not working (Coe, 2022). The Bill was praised by some MPs for tackling concerns about the internet, increasingly perceived akin to a 'Wild West online' (BBC, 2021). The lack of content regulation had left users, and vulnerable persons in particular, open to abuse, violence and fraud. At the same time, digital service providers were seen as benefiting from the perceived Wild West, as it allowed them to operate without impunity and develop business models that placed profit over safety. That said, the Bill was not without its detractors, with some digital rights campaigners outlining concerns over the overarching governmental powers that could give rise to state-backed censorship (Burns, 2021a; Smith, 2020).

It is in this rather febrile atmosphere that the OSA began its legislative passage, which itself was protracted (McGlynn et al, 2024). Prior to the OSA, the UK Government held two rounds of public consultation in 2017 and 2019 which led to the development of the initial Online Harms Bill, and then the subsequent Online Safety Bill, in 2022, which eventually led to the OSA (Nash and Felton, 2024). The change in focus from harms to safety is significant as it belies a change in government policy about who should be responsible for the prevention of online harms and how it should be done (Nash, 2019). The change represents a shift from *ex post* allocation of responsibility based on harmful content to an *ex-ante* approach of safety. The new approach goes beyond solely harm-based content to consider the potential impacts of business models more broadly. The systems-based perspective at the heart of the OSA gives rise to significant responsibilities for digital service providers and allows the state great scrutiny powers to examine and shape business model development (Nash and Felton, 2024). Moreover, the OSA's development also encountered troubled political waters that involved a backbench revolt and significant amendments proposed by the House of Lords which eventually shaped the OSA's final construction (Nash and Felton, 2024).

Unsurprisingly, the nature of harm underpinning online harms was an area of contention. The Online Harms White Paper of 2019 covered a panoply of harmful online situations ranging from national security to individual harms. The breadth of potentially harmful situations meant that the basis of harm relevant to government actions was unclear. Imposing a duty of care relating to harmful situations online based on content, when the harm is manifest and unclear, was challenging (Woods, 2019). A different regulatory approach was required and the 'starting point' of harmful content was wrong (Perrin and Woods, 2018). Regulatory action should instead focus on the broader online environment created by service providers and what providers could do to make that environment safer. The environments created by online service providers were viewed as public spaces and needed to be regulated with the same expected duties (Leiser and Harbinja, 2020). The greater focus on the online environment through a systems-based approach was beneficial because it better aligned responsibilities (Nash and Felton, 2024). This type of public space governance means the online spaces provided by service providers are not seen as technologically neutral. Instead, providers can use these spaces to shape user activities and actions, including for the profit motives of the provider.

The increased regulatory focus on environments, including business models, could better outline the risks arising from the design of digital services. It would also better align provider responsibilities with potential individual harms to ensure providers were responsible for ameliorating risks, especially to vulnerable users. Online providers would therefore be made more accountable for the potentially harmful content posted on their services (McGlynn et al, 2024) which would lead to a more positive digital experience for everyone (Nash and Felton, 2024). Greater accountability would create clearer processes of control relating to flows of harmful content. Online service providers would therefore be responsible for the spaces that they govern and manage the hazards generated within their space, including from how business models are constructed (Price, 2021). Doing so places responsibility on online service providers to implement greater protective measures as part of their core business models and operations.

The new regulatory focus was based on health and safety regulation which had a much greater focus on business model design and the core idea that providers have a moral responsibility to consider potential harms that may arise to users and to actively mitigate against them (Bearitz and Schertel Mendes, 2023). The imposition of a statutory duty of care thus involved proactive management of defined risks arising from providers rather than a universal duty to not do harm (Woods, 2019). The regulatory duties required by different service providers were thus differentiated and reflected the risks arising from individual services and their degrees of scale that could extenuate harms for users (Tambini, 2019).

The OSA identifies distinct types of service provider. Category 1 providers that entail high risk, high impact for users, such as pornographic services, have the highest obligations. Category 2 providers are high reach services, such as major search engines, or providers that entail higher

Figure 3.1: A new regulatory approach

risk in their function or operation, because a certain type of data is a core part of their business model. Accordingly, the duty of care imposed is differentiated because the first duty is based on provider and the second is based on content (Tambini, 2019). The types of legal responsibility that arise thus depend on the type of provider, their scale and content that is generated on their platform. That said, all providers have duties relating to the prevention and removal of illegal content and/or content that could be harmful to children. The child safety requirement gave rise to a greater focus on age verification which became one of the key contentions outlined in the bill passage debates.

The Bill thus established a duty to protect in different ways and for different types of provider audience. Three duties were initially proposed (Trengove et al, 2022) to: (1) protect all users from illegal content, (2) provide additional protective measures to keep children safe, if the service was likely to be used by children, and (3) protect all users from content that is harmful without being illegal, if the service was of sufficient reach and magnitude. The latter proved particularly controversial and was not included in the final draft of the OSA. Nevertheless, the OSA, as implemented, features a significant degree of the Bill's components.

The OSA regulates providers by establishing systems-based responsibilities to ensure positive digital experiences, especially for children. Service providers, under these duties, 'bear a preventative duty of care to users in the design and delivery of online services' (Nash, 2019). The OSA imposes duties of care on both types of provider in relation to two broad types of duties: (1) to conduct risk assessments based on potentially harmful content and (2) to mitigate potential risks arising. Consequently, even though the OSA takes a broader environmental approach, it nonetheless still retains aspects of its initial focus on harmful content, except now as part of the broader imposition of duty-based statutory responsibilities. Risk assessment is thus a key part of the responsibilities expected. Risk assessments regarding illegal content are required by the OSA for service providers to factor in freedom of expression and privacy. Services provided for children have additional responsibilities for risk assessments, to protect users from harmful content (Nash, 2019). The OSA also has extra jurisdictional application that covers service providers with links to the UK, that has UK users as part of a target market and which can include a service with a significant number of users or a service available in the UK that may give rise to a material risk of significant harm to individuals in the UK (McGlynn et al, 2024).

Taking on board these points, the OSA requires regulatory input which is primarily provided by Ofcom. Ofcom is tasked under the OSA to provide guidance on risk assessments and devise codes of practice in relation to the application of duties. Compliance with Ofcom derived codes fulfils the OSA safety duties for regulated service providers (McGlynn et al, 2024).

Ofcom can also monitor compliance with codes of practice and can enforce alternative measures to better meet Ofcom guidance, if a service provider is deemed to not be meeting expectations. Ofcom's strongest powers regard investigation and enforcement where it can issue corrective orders and impose fines if a code breach is demonstrated. Ofcom can fine companies up to £18 million, or 10% of annual global turnover, whichever is higher (Coe, 2022). In that sense, even though the OSA is in effect a legislative command, it is nonetheless representative of a hybrid regulatory approach which blends elements of public and private governance (Bearitz and Schertel Mendes, 2023).

In summary, the OSA's focus on online service providers' design and operation reflects a greater focus on business model construction and application as a key cause and generator of harms. The Online Safety Bill gave rise to a new systems-based approach to online regulation of online harms that does not focus on the removal of a particular type of content per se. Regulatory response and compliance activity is more about a provider's underpinning infrastructure and operations that gives rise to systemic issues which could have greater impact through amplification of content, especially through advertising-based business models. The key consideration that links to business model application is a direct recognition that certain types of business action are designed to influence user behaviour. Online service provider business models and their operation are not technologically neutral. The OSA pre-legislative journey recognised that they were designed to achieve a desired outcome and that outcome can have a negative and harmful impact on users and the broader societal environment. As such, the nature of harm that underpins the OSA's application shifted from one that focused on individual harms to one that recognised the societal responsibilities that online providers should have to create and develop online spaces safely (Trengove et al, 2022). However, the shift to a broader and more abstract notion of harm generation, with its focus on environmental space and community dimension, means that the ascription of individual responsibility, including to service providers, is potentially more challenging, as outlined in the key debates leading up to the Bill's passing.

Research approach

To identify the responsibility contentions that arose from fractious pre-legislative debates, we examined the four key stages and legislative components of the Bill. They were:

1. The Online Harms White Paper – published by the UK Government in April 2019, the White Paper sets out the government's initial proposals for the Bill (Department for Digital, Culture, Media & Sport and Home Office, 2019). Resulting from commitments made in the Internet Safety Strategy Green Paper (Department for Digital, Culture, Media & Sport, 2017), the White Paper also includes a public consultation.
2. The draft Online Safety Bill – published by the UK Government in May 2021.
3. We also draw on expert oral evidence provided during the Joint Committee on the Draft Online Safety Bill between September and November 2021 (referenced in the paper as 'JC#').
4. The Online Safety Bill itself, which was published by the UK Government in March 2022.

Documents were coded thematically to identify underlying themes and issues relevant to responsibility actions and concomitant contentions. Our initial analysis of relevant documents highlighted two issues of contention arising from pre-legislative debates: age verification requirements and scam advertising.

We focused on these two debates in greater depth as they highlighted significant discussion and contention about the need for online service provider actions and responsibility for both concerns and ameliorative processes. Equally important, both concerns highlighted debate about how digital responsibility can manifest in private sector product design. We outline both concerns in more depth.

Debate#1 – age verification

One of the 'landmark' proposals from the Bill aimed to tackle underage access to pornography. Relied-upon research demonstrated that approximately 51% of children aged 11–13 had seen pornography (Department for Digital, Culture, Media & Sport, 2022). Children's charities also expressed concerns about the damaging impact pornography can have on children's health, behaviour, relationships and the risks associated with child abuse and exploitation (Barnardo's, 2021). The Bill mandated all pornography websites, as well as platforms that publish and place pornographic material on their sites, to introduce age verification technologies and processes to access for users under the age of 18, replicating the policy direction relating to online gambling. The proposal was also in line with protections outlined in the Digital Economy Act 2017 and the Information Commissioner's Office (ICO) Age Appropriate Design Code (2020) which is a code of practice for online platforms providing services relevant to children. Verification checks would not require a full identity check and, by ensuring such technologies are 'secure, effective and privacy-preserving' (Department for Digital, Culture, Media & Sport, 2022), they should not process/store any data that is not relevant for the purpose of age verification.

The Bill's age verification proposals were largely welcomed, particularly by children's rights groups, but they also sparked ancillary debates over the scope of the legislation and the powers it affords. For some, the proposal threatened 'the integrity of the Internet's architecture' (Burns, 2021b) while others contended that the Bill's scope did not go far enough.

For example, children's charity Barnardo's warned that the Bill's categorisation of online service providers could also 'lead to a two-tier system that means children [could be] potentially able to access harmful content through these smaller sites' (Barnardo's, 2021).

The new focus on age verification, and the technological controls required to implement this, raised issues about the distribution of responsibility that led to fractious debate between private sector providers, government and civil society. The proposal was perceived as a shift of responsibility from the state to online service providers to provide age verification services as part of their core business model. The proposed new allocation of age verification responsibilities onto a new, emerging market was one that had been sought by civil society activists over previous years (Burns, 2021b). While this new form of responsibility allocation was set in train by the Bill, it was not absolute because 'the onus will be on the companies themselves to decide how to comply with their new legal duty. The bill … is flexible to allow for innovation and the development and use of more effective technology in the future' (Department for Digital, Culture, Media & Sport, 2022). Online service providers were therefore given leeway to select technological solutions appropriate to their circumstances including the impact on their overall business model. The delegation of technical solution to online service providers was criticised as it would allow providers to choose age verification controls that are more favourable to their business model and approach. The allocation of age verification responsibility and the degree it should be actioned by providers was a key element of this fractious debate.

Debate#2 – scam advertising

The Bill also proposed new solutions to tackle scam ads or online ad fraud that had become 'endemic' over the previous decade (Cookson, 2016). Ad fraud and scam ads grew dramatically during the COVID-19 pandemic, with digital marketers reporting billions of losses in 2020 alone. One of the key areas of criminal activity utilising 'paid-for' advertising business models involved targeted advertisements relating to deceptive products/financial schemes that involved well-known, trusted and reputable brands or individuals (Angeloni, 2022). Several studies demonstrated that UK consumers felt exposed and powerless in tackling scam ads and believed that online providers were not doing enough to support and protect them from harmful content online – a view that was shared by several expert witness testimonies during the online harms debate in Parliament (JC-1, JC-2, JC-3, JC-4, JC-9). Prior to the introduction of the Bill, the state appears to have taken on the burden of responsibility in this area, as the UK Advertising Standards Authority (ASA), the national advertising regulator, established its own Scam Ad Alert system. Introduced in 2020, the system provided a dedicated site/portal for UK consumers to report scam ads when spotted online, which would then lead to further cooperation with technology companies to take down scam ads (Advertising Standards Authority, 2020).

The building of scam ad mitigation into the Bill led to a fractious debate about the reallocation of responsibilities between the state, online service providers and individual citizens. The debate mirrored many others in the pre-legislative phase because it involved core issues of responsibility relating to actions required to mitigate harm. Ultimately, if the state or the private sector online service providers were unable or unwilling to block scam ads, then the consumer becomes the last line of defence for identifying and mitigating against harmful content – a requirement that would go against the core purpose of protecting individuals better against online harms that were difficult to detect by users. As such, allocating responsibility to end users would mean that individuals were deemed to have the capacity and capability to both identify and report scam ads, when there was a raft of evidence presented to the Committee that clearly indicated that was

Figure 3.2: Multiple actors with responsibilities

not the case. One of the key contentions arising from debate was that, despite the widespread harms scam ads cause, the business models deployed by the providers have paid-for advertisements at the core of service provision. Consequently, paid-for advertisements bring revenue to providers, whereas targeting harmful content, such as scam ads, is often regarded by providers as a regulatory cost. The Bill therefore sought to provide state influence by reallocating greater responsibilities to online service providers, while delegating a degree of latitude to preserve existing business models. The shifting responsibility for scam ad protection was reallocated from consumers to providers and it is this reallocation that gave rise to fractious debate and responsibility contention.

Findings

Our analysis of the pre-legislative debates revealed two core themes: (1) the responsibility contentions regarding the role of the state, online service providers and individuals regarding responsibility actions required to make the online space safer and (2) how digital responsibility manifests in private sector product design. Each theme is considered in greater detail next.

Responsibility contentions

The contentions that arise from the Bill debates analysis shows that there were different perspectives about which party bears responsibility for the implementation of technological systems to verify age of certain users and to prevent scam ads. The contentions are representative of the different responsibility actions required to implement solutions and the disagreements that flow regarding the allocation and discharge of responsibility. Responsibility contentions emerged in several thematic debates ranging from party willingness to absorb or avoid responsibility.

Throughout the debates, we identified situations where different parties were willing to accept or absorb responsibility for certain required actions and where the same or other parties sought to refuse or avoid responsibility for the technological implementation. As noted earlier, the debates highlight continuing attempts by the state and civil society groups for the Online Safety Bill to make providers more responsible for the publication of potentially harmful content on their platforms or for tackling online harms more broadly. The main legal vehicle for imposing greater responsibility through the Bill was the imposition of a general duty on providers to protect users from harmful content. Doing so would give rise to changing roles and relationships for state/regulator responsibilities and reducing the operational strategies of providers which placed greater responsibility for identifying and removing harmful content on individuals/consumers.

A number of expert witnesses and MPs/Lords argued that tackling harmful content was not a priority for providers, and that they were effectively abdicating responsibility for the

Figure 3.3: Putting profit first

safety of their users, even in situations where content breached their own rules and guidelines. For example, Imran Ahmed, founder and CEO of the Center for Countering Digital Hate, suggested that companies have their own 'playbook' in avoiding responsibility and were thus 'incapable of regulating themselves' and 'put profit before people' (JC-1). When asked why a lot of companies don't do more work to target/remove harmful content, Ahmed suggested that if they did, 'they would have moral culpability' (JC-1). That said, even though it was identified that providers avoided responsibility, it was often unclear where and how responsibilities should be located. A panel of regulatory representatives from key regulators involved in harmful content moderation (Financial Conduct Authority (FCA), ASA and Competitions and Marketing Authority (CMA)) provided evidence to MPs that it was not clear where regulator and provider responsibilities should be allocated. It was not surprising, therefore, that responsibility allocation was a contentious issue.

From the conflicts and tensions outlined so far, it is possible to identify a vacuum existing in the debates where several stakeholders were not entirely sure, or disagree, on where certain responsibilities should lie and with whom. These 'responsibility gaps' consisted of both legal considerations and societal/moralistic gaps, especially in relation to the actions of providers. In a legal/technical sense, contentions centred on the lack of regulation and corresponding security or safety controls that allowed providers to operate with relative impunity. On the latter point, one debate discussion noted how the Bill and online harms debate had highlighted a notable legislative gap for online content, especially in comparison to the way other media (such as TV or radio) are regulated (JC-9). For Clare Pelham, Chief Executive of the Epilepsy Society, the difference accentuated the isolation that many users experience online and how providers are 'completely beyond the reach of the law' (JC-4). Speaking to the claim that social media has become the 'Wild West online', these types of views demonstrate that there is little to zero accountability for online harm perpetrators operating on platforms nor the tech companies that allow perpetrators to function. In other words, everyone was in '[S]ome sort of badlands where there is no criminal law, no protection, no possibility of prosecution' (JC-4).

There was also clear contention and a responsibility gap about the protection of users' rights to freedom of expression and privacy in tackling harmful content online. This contention is of course not new and was a major consideration throughout the online harms debate. Concerns were raised about the Bill and the sweeping powers and responsibilities it gave to the state through Ofcom and the threat new regulatory powers posed to individual freedoms and rights, particularly in online spaces. The Joint Committee noted the Draft Bill 'provides duties on providers (and Ofcom) to have regard to the importance of the right to freedom and expression and privacy in various ways' (UK Parliament, 2022). While the Bill was supported by several organisations, particularly in the child protection community, several civil society organisations expressed major concerns about the Bill's scope and reach. The Internet Society claimed that 'the UK government was trying to legislate the impossible – a safe Internet without strong encryption – in the face of clear and consistent technical analysis saying that it cannot be done' (Internet Society, 2021). Similarly, Silkie Carlo, Director of Big Brother Watch, expressed concern during her Joint Committee testimony about the overarching regulatory powers that are tantamount to 'state-backed censorship of lawful communications' (JC-12). During her testimony, Carlo was challenged on this point by the Chair of the Joint Committee, Damian Collins MP. Collins took a different tack. Rather than solely focusing on the issue of whether content was unlawful or not, such as content that glamorises self-harm or anorexia under the UK's Equality Act, the more pertinent issue regarded 'to what extent companies are responsible for the environment they create and the harm that it could do' (JC-12). The role of the state and the protection of basic rights and freedoms was in itself unclear from a responsibility perspective.

Another key responsibility tension in the debates involved the notion of responsibilisation, whereby an individual (or group in society) is rendered responsible for a task or duty that previously would have been the role/responsibility of, say, the state or an associated institution.

In the context of the Bill and the wider online safety debate, responsibilisation is most contentious in relation to the expected role and responsibility of users. Responsibilisation contentions often surfaced in debates about incident reporting and the sense that citizens were being unduly relied on to report harms they encounter online. Individual users were thus being asked to bear the role of content moderation, reporting and policing that should be expected by both the state and providers (JC-9). To make individual users responsible for content moderation placed individuals in the situation of having the responsibility

Figure 3.4: The impact of responsibilisation on users

of monitoring, reporting and protecting themselves from scam ads knowingly or unknowingly facilitated by provider business models and systems.

Related to responsibilisation, the issue of improving user digital/media literacy was also a persistent contention. Enhancing digital literacy of users will provide stronger self-protection mechanisms and will provide better choice mechanisms for users to identify and to not engage in disseminating harmful content online. Educative responsibility is partly borne by the state as it has been ascribed to Ofcom, though it still needs to be acknowledged that while the state carries some responsibility it still requires a significant requirement by individual users to (a) engage with such education, but also (b) put their improved literacy into practice.

Digital responsibility manifestations in product design

Running adjacent to contests about responsibilities and their allocation to ensure for online safety was an increased focus by the state on the business models of online service providers. To a certain extent, the greater focus on provider business models is not surprising given the whole policy purpose of law reform involved a recognition that online spaces needed to be regulated differently. Unsurprisingly, then, Joint Committee debates highlighted separations and differences between digital content as a potential source of harm to consumers and the requirement for the design, management and implementation of technology to ameliorate and protect consumers from such harms.

Like individual actions based on responsibilisation, the role of content moderation regarding information that could constitute 'harmful' was often used to shape discussions and surface tensions about responsibility allocation. Principally, debates about this issue concerned the targeted removal or takedown of harmful content, which has traditionally been the responsibility of online service providers who are often criticised for not taking enough action. A contentious part of the Bill debates regarded whether the regulator, Ofcom, should have more power to mandate companies to identify and remove content it deems to be harmful. Thus, the shifting role of the state emanated through the Bill's development. During Joint Committee witness testimonies, concerns were expressed about delegating too much implementation leeway to providers while other respondents believed that Parliament should set prescriptive requirements in legislation. The focus on potentially harmful content, and its management by providers, was therefore a specific source of consideration regarding product design. As noted earlier, the regulatory focus on product design as a core safety measure is one of the OSA's key features and one of its most controversial as it involves a change in responsibility by the state involving willingness to influence and shape market forces.

An ancillary debate to content management also emerged about how provider technologies are designed, managed and implemented in different contexts relevant to harmful content and business models. Moving beyond issues around the scope and effectiveness of content moderation, such debates focused on the various systems, processes and design features that have led to the creation and amplification of harmful content and activity online. Algorithmic accountability and platform recommendation systems were key considerations, and how such systems were designed to create profit through advertising and by keeping consumers online through strategies to attract and retain cognitive attention. Again, the state's willingness to consider this degree of involvement in shaping business model construction, and the actions that flow from core business operations, clearly signals a change in responsibility role and the actions required to mitigate broader, societal harms.

Contentious debates focused on whether and what responsibilities should be placed on service provider designers to ensure that they are made more responsible and accountable for the design and functionality of their platforms and the harms they could potentially create. Consequently, many of the most fractious elements of the debate related to the business models of providers and the potentially corrosive relationship between harmful content, consumers and profit motives. The debates highlight the fluctuations involving the different realms in which digital responsibilities are formed, namely the market, the state, the regulator and the individual.

Figure 3.5: Pre-set responsibilities in business models

Debates about the broader social consequences of provider business models became a key backdrop in which the role of the state and consumer with that of the structure of the market and the digital infrastructure that gives rise to online service provision were often played out. A key focus regarded the perceived imbalance for monetisation and profit at the expense of the safety and security of consumers. Provider business models were perceived as the root cause for many problems as they present themselves as 'free of charge' when they were really 'rooted in targeted advertising' based on maximising attention (JC-4). The crux of the issue is that, although online service providers have the technologies and capability to deal with the takedown/removal of most harmful content on their platforms (and already did so to a relative extent), it is argued by many commentators that they took a lax approach to tackling such content. Doing so may impact upon profits either through potential loss of advertising revenue or human resources costs to implement an effective moderation system.

In the Joint Committee hearings, this view was shared by MPs, academics, technologists and experts from charities/campaign groups, who would often question what is stopping providers from doing more to tackle online harms and whether they were effectively 'profiteering from prejudice' (JC-2). Jimmy Wales (JC-6), the founder of Wikipedia, offered in his testimony a different business model that encompassed shared responsibilities across the platform. Under Wikipedia's business model, it was all users' responsibility to identify and remove misinformation. Content moderation was thus a shared responsibility of service provision and maintenance rather than a cost or threat to advertising-based business models. The juxtaposition of these two opposing business models reveals how content moderation responsibilities are realised and actioned to respond to online harms.

The criticism is well known to providers because it was previously used to highlight the lack of responsibility taken in relation to several key contentious activities involving potentially harmful content, including the proliferation of hate-speech, the live-streaming of terrorist content, algorithms promoting/linking to suicide content or the prevalence of scam ads that target vulnerable individuals. The Bill was consequently one of the first attempts to scrutinise and shape the dichotomy between provider made for profit business models and the safety of users. In doing so, the Bill would not only shape the business models of providers but also the responsibilities shaped the overarching digital market economy itself. Throughout the Bill development process, the question of whether providers would change their business models, and Parliament's ability to make them do so, was consistently ventilated. The Bill, in that sense, could set a precedent and send a message that the state was ready to govern the construction and operation of business model application.

Nevertheless, scepticism remained, with some expert witnesses suggesting that the Bill should go further and make providers, and company executives of providers, criminally liable for harmful content not removed from their platforms. The imposition of criminal measures would ascribe 'personal responsibility for senior managers in social media companies' and 'everything else will flow from it' (JC-4).

Thinking about digital responsibility

Our analysis indicates that the establishment and actioning of digital responsibility should not be assumed as axiomatic. During the debates, a responsibility consensus across key parties was not fully realised which meant that legal responsibility was established through the implementation of the OSA which commanded responsible action discharged through largely technological means rather than it being a consensus driven exercise. The Parliamentary debates thus demonstrate diverse types of responsibility action at play, which were mostly negative, particularly from the private sector parties, in trying to avoid responsibility for online harms and safety.

The debate contentions highlight that the to and fro of responsibility allocation is not unidirectional. Even though the OSA seems to indicate that once legal responsibility is established, through a Bill successfully completing its pre-legislative process and becoming a statute, it nonetheless still provides online service providers with avenues of interpretive leeway. Unidirectional application is thus not absolute, at least not in this situation. On its face, the pre-legislative debate history of the OSA presents an example of governmental activity taken where the private sector was unwilling to take responsibility for actions to create a safer environment by effectively mitigating potential content harms. The history, in one sense, could be viewed through this singular lens as purely an outcome of state action against unwilling regulatee response. Doing so, however, does not tell the full story. Instead, we contend that digital responsibility is multi-directional. The establishment and actioning of deemed responsible outcomes flowed in diverse directions. Some predictably flows from a command structure, such as the attempts by the state and civil society groups to make the private sector more responsible for their business models. Equally, on its face, the predictable actions taken by online service providers to avoid responsibility being allocated to them means that certain actions, such as reporting harmful content, become individual user responsibilities supported through technical processes and increasing educational capabilities. In this sense, the state also imposes some degree of responsibility on users by expecting them to educate themselves as a form of protection, especially from scam ads.

However, the process of establishing responsibility, and the actions produced, is dynamic rather than static. Take, for example, the civil society criticism regarding the OSA's potential for state-sponsored censorship. Consensus alignment takes place at various times, for example, when the state and civil society is aligned in critique of service provider business models. Yet, civil society and the state can be mis-aligned,

Figure 3.6: The flow of responsibility is multi-directional

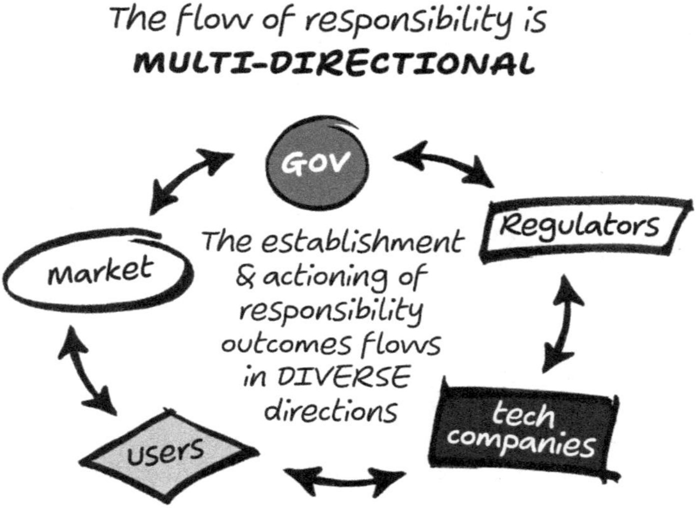

such as regarding the need for protection of fundamental rights or censorship concerns. Equally, the state and service providers find consensus alignment at certain points of the ongoing debate, such as the imposition of criminal penalties for provider senior managers, even though the proposal was largely supported by civil society. Again, these differences can be viewed as the normal cut and thrust of pre-legislative debate in which differences can be attributed to different notions of issue importance. Through a digital responsibility lens, the alignments and mis-alignments point to the different points of consensus that are required and generated when responsibility allocation is under discussion. Understanding the location points of consensus and recognising the actions that flow from established positions of alignment and mis-alignment are consequently important.

Our analysis of the debates showed the establishment of the OSA's differentiated duties of care, disrupted responsibility actions and expectations of which party should undertake desired regulatory actions. As noted earlier, even expert regulators with knowledge and experience of harms in the online sphere were concerned about where responsibilities could be located, including how state/provider responsibilities should be allocated. The situation is further complicated in an environment where the state attempts to impose responsibility, and the providers who would be imposed upon attempt to avoid responsibility by allocating actions and requirements to users. All of this shows that digital responsibility is transitory and flexible, which goes to explain why responsibility establishment and allocation of action are contentious issues.

The idea that responsibility establishment and actioning have a location consideration is important because it speaks to the importance of gaps, as highlighted earlier. The location of many debate contentions was the business model of service providers, the site where responsibility should be allocated and actioned but was not. Location is consequently important to consider because it highlights that responsibility is formed across locations and environments, which points to the multi-directional application of responsibility in the digital context.

In the OSA debates, the focus of business model contentions indicated that multiple parties thought regulation needed to be done differently and thus content moderation became the location focus of where regulatory activity needed to take place and the focus of responsibility allocation in terms of who should take reporting action, particularly the provider or user. The same could also be said regarding the regulatory gap and uncertain location involving the application of security controls. The imposition of controls was believed to be beyond the effective reach of law and responsibility establishment was therefore about the choice of appropriate technological measures which was best left to provider knowledge of needs. This also has much wider implications for the effectiveness of the security controls that form part of the digital design, for if the business model itself does not contain a framework that protects end users/consumers, then security controls will be less able to protect the end user/consumer. Unless the security goals for which controls are implemented are supported by an agreed distribution of responsibilities in the business model that promotes collaboration between the parties, the security controls will be limited in their effectiveness.

The idea of location and gaps also has a different dimension in the responsibility context. Responsibility gaps indicate consensus mis-alignments where responsibility allocation or actions fall into a lacuna and no party is responsible. The gaps themselves are multifaceted and can cross different environmental boundaries, such as public/private or legal/moral, which further diffuses the location and allocation of responsibility. Similarly, as highlighted earlier, a responsibilisation gap signifies situations where the state has retreated from undertaking certain responsibilities, thus creating a gap that needs to be filled. The latter highlights the shifting role of the state, including in the OSA debates. The state retreated from the management of online spaces and left

Figure 3.7: The sites of the state, the market and the individual

that activity to the market to establish responsibilities through individual terms of service. Yet, through the OSA, the state has returned to the construction of management expectations regarding safe online spaces and directly attempts to influence and shape market forces to create a safer environment by the prevention of harm. The OSA's focus on a safe online environment creates changing roles that the state is willing to establish responsibility and changing actions to ensure that establishment is fulfilled. Again, responsibility flows multi-directionally as state establishment and re-establishment flows across different realms and the state creates new ways to augment existing responsibility actions, such as takedown notices. Finally, the debates highlight a desire for shared responsibility. We read this desire as a signal for consensus where a gap has emerged. Accordingly, content moderation is not just to the action of taking down harmful content and it is more about the agreement of all parties to monitor and regulate across the digital environment.

Conclusion

Our analysis demonstrates how digital responsibility is established and realised at state-level. It highlights the digital policy challenge of tackling online harms as a prism to (a) examine the fractious policy debates to identify the responsibility contentions behind the debates and (b) identify how digital responsibility manifests itself within the different dimensions of digital product design, delivery and use by the private sector. Our overview of the online harms debate concluded that the business model of tech companies (and its relationship to the wider market economy) has pre-set the conditions for responsibility in this context. In other words, responsibilities in the context of the online harms debate are very much dictated and led by the business model of online service providers. For example, when considering how certain technologies are designed, managed and implemented on platforms, the conditionality shaping digital responsibility (be it a 'duty of care' or the protection of freedom of speech) is intrinsically linked to the market and business model of the platform. A few different tensions and conflicts arose between stakeholders which highlights that establishing responsibilities, even through state commanded forms of legislative action, such as the OSA, will not in itself obviate the need for consensus among core stakeholders.

FOUR

The Digital Turn in Migration

Evan Easton-Calabria

Introduction

In this chapter, digital responsibilities are considered through an international lens, considering how the deployment of digital technologies by states shapes the nature of responsibility in the context of refugees and migration. This study therefore sits at the intersection of humanitarian responsibility and digital responsibility.

When one thinks of responsibility and refugees, the obligation for states to host refugees, such as adherence to the 1951 Refugee Convention, might first come to mind. Indeed, the United Nations' (UN) 2018 Global Compact on Refugees – the current global guiding document on refugees – laid out a vision for 'more predictable and equitable burden and responsibility-sharing' for refugee situations (United Nations, 2018). It sought to 'operationalize the principle of burden and responsibility-sharing, to mobilize the international community as a whole, and to galvanize action for an improved response to refugee situations' (United Nations, 2018, p III). This vision is laudable and important: over 122 million people are currently forcibly displaced (UNHCR, 2023). Less than 1% will ever be resettled to a third country and cross-country agreements for refugee resettlement are gaining traction and growing voter support (Vrânceanu et al, 2023). Such policies

Figure 4.1: UN Global Compact

are dependent on a digital infrastructure that can administrate the cross-country movement and resettlement of refugees. This infrastructure brings with it new forms of responsibility that highlight the antecedents from how responsibility actions emerge, how they are mediated and mitigated by digital technologies.

While concerning itself with the notion of responsibility in refugee situations, this chapter narrows the focus to a similarly under-explored concept, that of digital responsibility. The issue of digital responsibility and refugees is scant in the academic literature, despite the growing focus on the importance of internet connectivity and digital administration in both transit and displacement since the 2015–16 so-called 'European refugee crisis'. In this chapter, we sketch the literature that frames the discussion on digital responsibility and refugees and present two case studies that animate the themes in the literature, focusing on the digital responsibility of humanitarian agencies, namely the UN Refugee Agency in global displacement, and the work of refugee-serving civil society organisations supporting the digital inclusion of refugees in London during the COVID-19 pandemic. We close with a discussion of how the findings align with the book's digital responsibility framework.

Background

Responsibility in the international refugee regime

While responsibility-sharing is commonly understood as the equitable sharing of both finances for refugee assistance and intake, and numbers of refugees resettled, in reality this is not practiced by States. Instead, the global norm of responsibility-sharing remains much more idealised than real, thereby precluding more nuanced discussions of what responsibility in refugee contexts actually entails. Both discussions and references to the notion of responsibility in the international refugee regime are complicated by the fact that the term 'responsibility' is rarely defined, with many texts simply relying on the assumption, based on the 1951 Refugee Convention, that States, either alone or collectively (Hurwitz, 2009), are legally responsible for taking care of them.

However, a small body of academic literature focuses directly on responsibility and refugees. Much of this literature is policy-oriented rather than critical, focusing on how to increase responsibility-sharing by States rather than interrogating broader notions of responsibility itself. Martin et al (2018) focus, for example, on areas of responsibility-sharing including addressing underlying causes of displacement, efforts to find solutions, initiatives to enhance protection, and technical assistance and training for host countries and organisations. Taking a legal perspective, Dowd and McAdam (2017, p 867) explore the 'two main methods of sharing responsibilities, namely the provision of financial and other assistance to host countries, and the admission of refugees' as well as the principle of common but differentiated responsibilities, exploring how it might apply in international refugee law. In a similar vein, Ahmed (2018) poses the critical question of who should take responsibility for so-called 'climate refugees', using a

Figure 4.2: Unclear responsibilities

model considering climate pollution, consumption, human development and other factors to suggest that Australia and the United States should take responsibility for 10% of climate refugees, followed by Canada and other countries. Furthermore, while there is limited understanding of the responsibilities related to refugees, there is even less about the need for digital needs and the responsibilities that flow.

Refugees and digital technologies

Since the large-scale arrival of Syrian refugees in Europe in 2015–16, myriad digital initiatives seeking to 'include' and facilitate everything from the transit to the integration of refugees have arisen. These developed in part out of a recognition that smartphones and connectivity were key tools for refugees in transit, helping them navigate, find safe places to sleep *en route* and access information about their destinations (Frouws et al, 2016; Gough and Gough, 2019). However, technology production is neither a neutral nor objective process but instead is shaped by geographies, ideologies, and multiple interests (Wahome and Graham, 2020). It then follows that the use and promotion of technology is similarly imbued with complexity. Despite the complexities inherent in digital service provision, refugees' digital inclusion is predominantly presented by governments and humanitarian and development agencies as a human right and an emancipatory tool with the potential to increase access to skills and income (UNHCR, 2022a).

A growing body of literature has also critically explored refugees' relationship to technology, as well as how humanitarian and development agencies act as mediators to digital inclusion. While not interrogating the notion of responsibility explicitly, Jacobsen and Sandvik (2018) explore UNHCR's

Figure 4.3: Digital assistance

international protection efforts and in particular their efforts to use 'accountability technologies', namely results-based management, biometrics and cash-based interventions. They argue that 'this configuration and use of accountability technologies contributes to the emergence of a particular understanding of international protection as a task best accomplished through improved techno-bureaucratic legibility and quantification practices', focusing on the role of technology in both enabling and obscuring particular practices related to 'accountability' and 'protection' (Jacobsen and Sandvik, 2018, p 1509).

Udwan et al (2020) developed a case study of Syrians living in the Netherlands and their 'digital resilience', focusing on digital social support, digital health and digital identities. They noted that media and digital networks have been overlooked for their positive role in fostering resilience, later sharing an excerpt from one Syrian vlogger's YouTube channel: 'I am the virtual daughter, I am the virtual wife, I am the virtual friend. I am the virtual sister. I am the virtual sweetheart. … In my alienation, I hug my mobile phone at night. Every morning, I give that metallic phone a virtual kiss' (Udwan et al, 2020, p 8). At the same time that they highlight how refugees employ technology in a variety of positive ways, they caution: 'While we have emphasized several digital resilience tactics refugees mobilize for self-support, health, and identity management, we want to avoid our results being co-opted by governments to demonstrate their policies of no-policies and retraction of assistance are successful' (Udwan et al, 2020, p 9).

These areas of facilitation include communication with the government, social connectedness, participation in educational programmes, and social inclusion (Krasnova et al, 2016). Digital technology is also used to create so-called 'digital work' projects for refugees, which often include integration as an aim. Alencar and Camargo (2023) describe a digital work initiative in Brazil that reflects the tensions between the aspirations and empowerment narratives that are woven across digital work and the barriers to access and the amplification of marginalisation that migrants can experience when undertaking work in the digital economy. AbuJarour et al (2017) posit that refugees encounter three major stakeholder groups within the bureaucratic and socio-economic structures of the asylum and resettlement process: (1) local government and public authorities, (2) local population and (3) businesses. In this view, digital technologies mediate or have the potential to increase engagement with these stakeholders.

However, other works highlight the security implications of such engagement, with a predominant focus on border and migration management (Ajana, 2013; Latonero and Kift, 2018), data privacy and usage, biometrics (Jacobsen and Sandvik, 2018; Završnik, 2021) and the concept of digital identities. Simko et al (2018) explore computer security and privacy for newly resettled refugees in the United States, finding that ICTs such as email are important in integrating into new roles while barriers to digital security are also paramount, including a lack of language and technical skills, and cultural knowledge, that

Figure 4.4: One size does not fit all

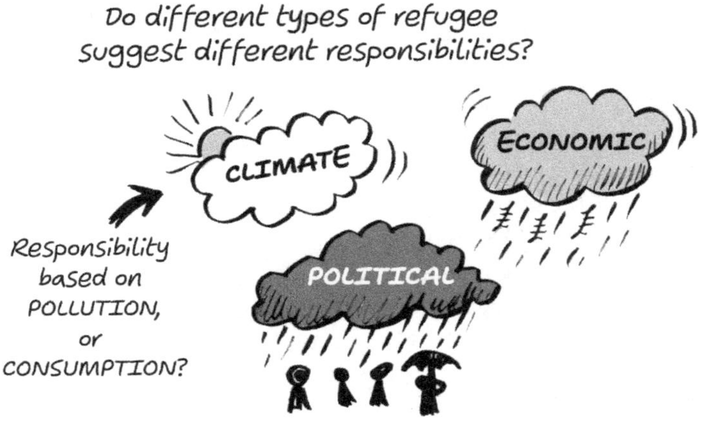

can lead to scams and other forms of digital exploitation. Other work explores, generally critically, the data-driven technologies of European border management such as EUROSUR, the European Border Surveillance Systems, and European-wide databases such as the Schengen Information System, finding that the erection of 'digital borders' and the 'European digital fortress' have – unsurprisingly – nefarious impacts on refugees and their rights (Topak, 2014).

Digital responsibilities and refugees

While rich and varied, much of this literature does not address the question of: Where does responsibility for refugees lie in these digital arenas?

The United Nations High Commissioner for Refugees (UNHCR) is mandated with the legal protection of refugees. While, at least on paper, host states are legally responsible for refugees, UNHCR has been called a 'surrogate state' (Deardorff Miller, 2017) and commonly steps in to support host states with functions such as asylum status determination and, through its implementing partners, everything from livelihoods to education. Increasingly since 2015, UNHCR has shown a heightened awareness of the importance of digital connectivity for refugees and some nascent efforts to connect refugees to digital work, complemented by a longer-term focus on biometric identification and other aspects of digital identity. Each of these arenas raises questions of how UNHCR addresses and indeed perceives its own digital responsibility towards refugees.

One of the most intriguing elements of digital responsibility in relation to UNHCR and refugees is the additional role of host states, as it is ultimately their policies and practices that define the legal and operational scope and context for humanitarian action. This extends to data management and other digital practices (Schoemaker et al, 2018). While UNHCR does offer support to connect refugees to the internet (expanded on in the following quote), it makes clear that digital inclusion is not solely its responsibility:

> Identity in the 21st century is no longer just paper-based and centred on breeder documents, such as birth certificates and ID cards. With new technologies providing access to the internet, mobile phones and related services, to information, education, banking, and other economic opportunities, the concept and realities of identity broadens. *States have the responsibility to provide for the digital inclusion and identity of their citizens and those living on their territory.* UNHCR assists member states in ensuring that refugees and asylum seekers, stateless persons, and other forcibly displaced are – digitally speaking – not left behind. (UNHCR, 2022b) [emphasis added]

Despite this position, UNHCR has steadily increased its engagement with digital inclusion. Some of this early work was piloted through the UNHCR Innovation Unit, established in 2012 to experiment and problem-solve within the agency. Focusing at the time on areas called Innovation Labs, the unit emphasised self-reliance in addition to ICT, access to energy, data and communication, and field delivery. One of these projects was Community Technology Access (CTA) in Kenya's Dadaab refugee camp (UNHCR, 2022c). This project has since expanded to both camps and urban areas and forms the practical basis for UNHCR's focus on helping refugees access digital work – what UNHCR originally termed teleworking activities – and wider digital skills.

It is notable that UNHCR has been involved with digital data such as biometric data collection for much longer than other efforts such as digital work for refugees. It has, for instance, used identity technology such as iris-recognition technology since 2002, when it was first employed for Afghan refugees returning from Pakistan (Jacobsen and Sandvik, 2018). What has vastly expanded, as well, is UNHCR and other humanitarian and development actors' focus on internet connectivity, which is often termed 'digital inclusivity' in policy documents and practitioner guidance. With this expansion has come a growing recognition of the risks and responsibility of acting as the mediator between refugees and internet connectivity, particularly in terms of actors such as the private sector that became collaborators (UNHCR, 2020a).

This more recent focus on digital responsibility – though rarely termed as such – largely comes from UNHCR's Digital Inclusion Programme, part of the Innovation Unit which brings together projects providing internet connectivity to refugees ('Connectivity for Refugees') with efforts to help refugees be better listened to within UNHCR through digital technology ('Communicating with Communities') (UNHCR, 2022a, 2022d). These strands of work bring together responsibility in multiple ways, with UNHCR presenting internet connectivity as an important right for refugees, that itself as a surrogate state has the responsibility to safely provide, and communicating with refugees as a humanitarian imperative that technology can enable. One 2020 UNHCR policy document evokes the concept of digital responsibility along with clear areas for institutional reflection and learning:

> Understanding how displaced communities find gateways to access the Internet, which factors influence and determine their choices, *what UNHCR' s mandate of protection means in a digital space*, or the extent to which specific technologies or tools can reduce or exacerbate inequalities, will inform and shape future efforts in providing connectivity to refugees in a safe, adapted, and dignified manner. (UNHCR, 2020b) [emphasis added]

Alongside this, a strong focus on digital identity has emerged. In 2017, UNHCR High Commissioner Filippo Grandi stated his aim for every refugee to hold a unique digital identity, stating that, 'This will enhance accountability and facilitate two-way communication between refugees and service providers' (UNHCR, 2018a). This has been echoed elsewhere in UNHCR as well, such as by UNHCR's Deputy Director of the Division of Programme Support and Management and backed up in programming: 'It is not only logical to promote the digital inclusion of refugees; in the digital age, *a digital identity for all is the only possible way* to make sure that no one is left behind' (UNHCR, 2017) [emphasis added].

This focus is intertwined with a rhetoric of empowerment for refugees. UNHCR posits that digital identity provides internet access, mobile phones and connected services, and that it is through this digital inclusion that 'empowerment passes through': 'Access to jobs, income and remittances, online learning and web-based economic activities will make a difference in the lives of people we care for' (UNHCR, 2017). There are a number of digital identity programmes used by UNHCR, including ProGres, PRIMES and SCOPE (UNHCR, 2018b). However, such programmes have been critiqued for not enabling refugees to access the services they need, and in the manner in which they need to access services (Ungar and Seymour, 2024). Moreover, such systems have been found to restrict interoperability with third-party tools and restrict choice as to how and when such identity systems are to be used.

Regarding data collection by humanitarian agencies more broadly, recent scholarship has identified key concerns including 'organisational data responsibility', defined as the ways different organisations understand (or don't understand) privacy rights and data security practices (Latonero et al, 2019). The growing use of digital and mobile technologies in the humanitarian sector poses a variety of risks for recipients such as refugees, including unauthorised third-party access that can lead to both data exploitation and indeed actual harm for people who have been identified (International Committee of the Red Cross, 2024); the risks more broadly of 'digital by default' has led to calls for a people-centred security practice that helps users become active participants in security learning (Coles-Kemp et al, 2020).

While digital responsibility in theory lies with organisations collecting data, some research finds that digital literacy and understandings of security management is severely lacking in many NGOs, particularly smaller and more grassroots ones. As one study focusing on Italy explained, 'The risks in the Italian situation stem from organizations that lack the knowledge to protect the information they collect or, if they possess that knowledge at the headquarters level, lack the ability to implement policy to those working locally' (Latonero et al, 2019, p 37). These researchers gave the example of visiting the office of a refugee-serving agency that claimed beneficiary data was secure but which in fact had an unsecured Wi-Fi network and transmitted data via a website lacking basic security/ encryption protocols (Latonero et al, 2019, p 37).

Although not commonly invoked in discussions on refugee assistance and protection, the concept of digital responsibility has the potential to be a useful tool as humanitarian organisations continue to grapple with when and how much data to collect, the need for digital literacy by both employees and refugees, how to improve accountability to affected populations through digital means, and how practices of protection and digital responsibility interrelate. If UNHCR and other humanitarian organisations are serious about protecting and upholding refugees' rights and increasing digital inclusion, then centring discussions on digital responsibility – their own and that of States – is an important next step.

Research approach and methods

Drawing on semi-structured qualitative interviews, desk-based research and participant observation of monthly meetings of a London-based civil society network focused on digital exclusion, this case study focuses on the 'in between' and mediating role of digital access and exclusion by refugee-serving civil society organisations in London.

The case study was carried out in the second year of the COVID-19 pandemic when UK society had become widely regarded as a digital-first society where essential and statutory services had become digital-default and where the digital was the foregrounded mode of access for the majority of services. It captures a moment in time when responsibilities emerging from, mitigated with and distributed by digital technologies were a point of discussion in the public discourse. It was also a point at which digital responsibilities with and for minoritised and underserved communities became recognised as an issue of public importance (Coles-Kemp and Hall, 2023). One such community is that of refugees and asylum seekers.

According to UNHCR, as of mid-2021, over 135,000 refugees, 83,000 pending asylum cases and almost 4,000 stateless people were in the UK (UNHCR, 2022e). Even before the start of the pandemic the number of refugees and asylum seekers in poverty in the UK was rising. Research shows

Figure 4.5: The COVID-19 context

that they are less likely to receive vital public health information due to language barriers and digital exclusion; they are also part of a minority population more at risk of severe illness and death from the virus. Barriers to public health information contributed to outbreaks in hostel accommodations in the UK and chaotic government responses to them (Faulkner and Lee, 2022). Refugees are also more likely to suffer from depression and PTSD than the local population. These challenges have been linked to both 'pre-migration experiences' (such as war) and 'post-migration conditions', such as inadequate housing and difficulties with asylum procedures (Porter and Haslam, 2005). The disruption to necessary welfare and health services that were experienced by the local population had therefore an amplified impact on refugee groups, thus making access to digital services vital.

As this study shows, voluntary and third sector organisations are both providers of digital access for refugee groups as well as often neglected actors in terms of government recognition and funding for digital inclusion. Digital responsibility lies both within these organisations as well as within actors such as the UK Government, which must provide further support in the form of digital literacy for refugee-serving organisations, and funding for digital technologies for both refugee-serving organisations and refugees themselves. The following sections present key themes raised in interviews and literature and their links to digital responsibility as it is enacted within the UK context of refugee resettlement.

Findings

The findings reflect three main themes that relate to responsibility and the roles digital technology played during the initial phase of the COVID-19 pandemic for the refugee groups taking part in this study. The first theme reflects the tensions that relate to the operationalisation of digital access in a digitally enabled society. The second theme highlights the digital responsibilities that were conferred upon voluntary and third sector organisations serving refugee groups and some of the challenges experienced by those organisations in actioning those responsibilities. The third theme considers the importance of trust if digital responsibilities are to be successfully actioned. These findings combine to paint a more nuanced picture of what happens when a State initiates the actioning of digital responsibilities without ensuring that the necessary policy and technology infrastructure is in place for individuals and groups to complete the actioning of those responsibilities. The case study also illustrates how voluntary and third sector organisations step in to absorb and co-ordinate the actioning of digital responsibilities. Above all these findings combine to underscore how difficult it becomes for digital responsibilities to be effectively actioned without a coherent framework that foregrounds and operationalises digital responsibilities.

The right to internet?

Although internet is increasingly considered a public good, with some cities having city-wide free public internet access, the pandemic has shown that this is not by any means always the case. In London, public spaces, like libraries and organisations where refugees and asylum seekers have typically been able to receive a variety of assistance and services such as free Wi-Fi, were shut during the pandemic. To add to this burden, the near overnight shift of services online as well as the digitalisation of normal everyday bureaucracy created a double challenge. Asylum seekers in the UK are not allowed to work and at the time were provided only £37.75 per week by the government. While many of them were previously able to make use of libraries, shopping centres and other public spaces for internet, during the pandemic they were often forced to shoulder these costs alone.

As one member of the UK Red Cross asked, 'Do you pay for phone data to speak to family or help your child access online classes? Or do you buy food?' One informant explained that some asylum seekers he worked with previously accessed Wi-Fi through a local community centre; when it closed during the pandemic, the only way they could get Wi-Fi was through a nearby bank, often just by standing outside it. 'Can you imagine?' he asked, 'Three or four young people standing outside of a bank for hours. This can seem suspicious, they may seem like they're causing trouble and are gangsters. People will wonder: why are they loitering outside of a bank?'

Although in June 2020 the UK Home Office began providing internet access via SIM cards for asylum seekers in certain accommodations based on Public Health England advice, no smartphones were provided to those that lacked them. During long stretches of the pandemic, social mixing was not allowed, so asylum seekers couldn't even go to each other's rooms in hostel accommodation to use internet through a friend. These and other situations illustrate significant gaps in helping asylum seekers access not only public health information but other crucial services, as well as make desperately needed social contact with family and friends. Even refugees and asylum seekers who do have smartphones often have no contract due to its cost, so instead use 'pay as you go' data as their only access to internet. One informant explained,

> In the past it might not have been so tricky to budget for data because you paid 5 pounds a month for it. But now it is 20, even 25 pounds. People are spending more than they expect. One lady I worked with used to spend 10 pounds per month but is now paying 35 pounds per week. Her language skills aren't very good and she didn't realise she could get a monthly package that would be cheaper.

Refugees and asylum seekers also may not be housed and thus deal with the added intersecting challenges of homelessness.

One informant who volunteers with refugees and asylum seekers as well as homeless populations in London stated,

> Parts of London look like camp site cities, areas completely covered with tents. In all these situations of homelessness you find a lot of immigrants and refugee populations. With the pandemic, homeless people are really struggling with access to services. For example, what happens when you don't have credit on your phone? People assume others have technology, but that's not the case. It has shocked me in terms of the discrimination people are facing.

The Department for Education's schemes to increase children's access to remote education during COVID-19 also extends to refugees and asylum seekers and includes increasing data allowances on mobile devices and providing 4G wireless routers. In this way, the responsibility for providing education was taken by the UK Government, with enabling digital access as the means to do so. However, there has been significant criticism that the scheme has not been extended far enough and that many were left out or under-served. Some informants also mentioned it being common for one family to share one device, such as a laptop, among themselves, with competing interests for school, job searches and social contact with family and friends leading to inadequate digital access.

And having some access to internet does not necessarily equate to having the *right* access. One organisation explained that several refugees they worked with were trying to gain job qualifications or get jobs but were stymied:

> In some instances clients wanted to attend CECS, which is a qualification to work on a construction site, or SAI training (security warden training), but there was a barrier there – they were unable to attend as they needed a laptop with access to Wi-Fi because of the nature of the training.

> Others have been asked to attend a job interview with a laptop – but they didn't have one so they couldn't attend the interview at all!

Such examples illustrate the importance of accounting for restricted or lack of digital access through both increased flexibility and increased provision of necessary technological devices. The evident risk of not doing so is allowing already marginalised populations to be left further behind.

The digital responsibilities of refugee-serving organisations

The shift to remote service provision matters for many reasons. An obvious one is language barriers that can be exacerbated by a lack of in-person assistance. Importantly, the challenge of addressing language barriers remotely is not just one that refugees and asylum seekers must deal with, but those civil society organisations working to assist them. These organisations appear to have both taken on the responsibility to foster digital literacy for refugees as well as had this responsibility in some cases foisted on to them, as refugees increasingly called on them for help navigating UK bureaucracy as it digitised during the pandemic.

Organisations described the initial challenge familiar to many of us of adapting to different technologies, such as launching zoom on a phone or downloading an app, but with the added struggle of explaining this to someone speaking a second, third or even fourth language. One organisation sought to overcome this by involving Arabic-speaking colleagues and using a tactic they referred to as 'nudging':

> We initially wanted to tap into the existing tech and skills that clients had, there was no point in insisting on them using something they didn't know how to use. But we did do some nudging. We asked them, why don't you try to touch this button or try a video call instead of a voice one?

Figure 4.6: Assuming responsibilities

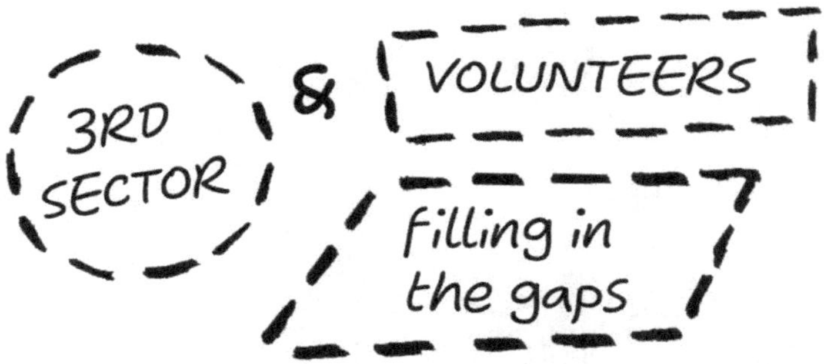

For refugees and asylum seekers themselves, a lack of language skills contributed to a fear of forms that either must be filled out on paper or are hard to access online without help. As one employee of a refugee-serving organisation explained, 'At the beginning, everyone was asking about the office being closed and when it would open again. People wanted help with paper forms – they were used to coming with forms for school registration, GP application, provisional driver's license, and so on.' Whereas previously clients could come to organisations or even sessions at public libraries to receive help with this bureaucracy, once the pandemic started everything had to happen remotely.

Yet some refugees and asylum seekers don't know how to write in English and might not have even gone to school in their home country. For these people, completing anything official or bureaucratic is overwhelming, and they need support for even so-called small undertakings like calling a water company to check about a bill. Many refugees, regardless of literacy, were also confused about changes to council benefits given changes in Universal Credit. This led to an increased reliance on help from organisations that refugees know and trust. One informant described a lengthy process of supporting a refugee who didn't speak English to receive clarification about

Figure 4.7: Bureaucratic complexity

apparent rent arrears; as everything was remote, the employee's colleague had to phone up the council with the client on the line, who had to confirm their identity and approve that someone else was allowed to speak on their behalf. It was only then that the actual issue could begin to be addressed.

There is also a broader issue which extends beyond refugees and asylum seekers: the possibility that some services cannot provide the same duty of care remotely as they can in person. This was mentioned in several interviews regarding social services like mental health support. There were concerns, for example, about the quality of mental health assessments over the phone, particularly for people for whom English is not their first language. This was brought up against a larger backdrop of declining services due to higher levels of demand, as well as the fact that only certain members of populations were accessing digital and remote services in the first place. As one employee of a charity working with refugees put it bluntly, 'We only [virtually] see the clients that are reaching us. We don't know about those who don't call.'

Digital responsibility and challenges of trust

Many of the challenges mentioned previously feed into a less tangible yet incredibly important one: that of trusting and gaining trust remotely. Several organisations interviewed discussed a drop in client numbers during the pandemic attributable not only to a lack of information but also to a fear by many refugees and asylum seekers of providing personal data online or over the phone (for example, through intake forms) to unknown and thus personally un-vetted organisations. It also was hard for organisations themselves, that had to quickly adapt to new working arrangements while attempting to retain access to clients. As one organisational employee stated, 'It was also a problem for us: how is this going to work? How are we going to communicate with these people and keep them involved in the project?'

Another employee of an organisation seeking to increase refugee employment explained,

> When a potential client used to come to our office, they saw people working there and others getting help. It looks like a charity, it's an actual physical environment – people can talk to someone with face-to-face contact. … When a potential client used to come to the office, they would figure out who they were dealing with, which increased their trust. So face-to-face is very important.

The shift to remote working, which meant that organisations in some cases could only communicate with refugees and asylum seekers over phones and text messages, was described as denying people the opportunity to have more natural human contact that fosters trust. While crucial for everyone, this is particularly important for forced migrants who often have a deep mistrust of institutions due to past experiences of violence and persecution and may also have significant language barriers that makes remote communication all the more daunting.

Thinking about digital responsibility

Throughout the pandemic, refugee-serving organisations in London took on digital responsibilities by supporting refugees with unforeseen tasks while also dealing with their own digital challenges, be it trying to run their own activities over Zoom or navigating new digital systems and databases for their work. In this way, charities and other civil society organisations, as well as individual volunteers, supporting refugees and asylum seekers in London and beyond played an enormous role in getting people the support they need. As one charity worker put it, 'The third sector helped massively because we were inundated with requests for support.'

While the responsibilities for supporting refugees have been established in law, and there is consensus about the need, these findings suggest there is no consensus about how responsibilities should be actioned in a society that is digital by default. This case study reveals how established responsibilities are not discharged because the actions are refused or avoided. It therefore shows that even if a responsibility is realised and established there still needs to be consensus on what it means for an individual to be response-able to action the responsibility. The study shows that this disconnect is especially profound where there is no enforcement and significant existing barriers to accessing digital services to action the response.

Refugees and asylum seekers in London and in the UK faced digital exclusion as both one challenge among many during the COVID-19 pandemic as well as an exacerbating factor in other challenges. A lack of or limited access to internet due to reasons such as prohibitive costs, lack of devices and lack of digital literacy all contributed to this population struggling to access the services they need. Language barriers remain significant and heavily impact the ability for many to navigate online bureaucracy without external support.

The case study explored here raises several particular elements of digital responsibility by humanitarian and civil society organisations towards refugees. At the core of these considerations is the notion of responsibility itself in refugee situations. As overviewed in the introduction, by its very definition, a refugee is no longer under the responsibility of their home country, but instead has a host state, UNHCR as a 'surrogate state' (in displacement) or a resettlement country which legally 'takes responsibility' for them. In this light, a lack of responsibility in the refugee context is one inherently related to the breakdown of the state-citizen contract or in fact relates to the lack of fulfilment of the (new) state-citizen contract. This in turn highlights refugee-serving organisations in the third sector as important actors in refugee assistance, which either explicitly or de facto take responsibility for various aspects of helping refugees. The contexts explored here illustrate that just as this responsibility occurs 'in real life' so does it manifest digitally, as well.

While more should be done to support refugee-serving and other civil society organisations in bridging the digital divide, many of the challenges refugees and asylum seekers have faced could in large part be overcome through the government either fulfilling or expanding its promises to provide devices and internet access to those that need them – in essence, by taking digital responsibility. In this context, digital responsibility means first and foremost taking concrete steps to enabling digital inclusion for refugees and asylum seekers, thereby inserting additional digital responsibilities before refugee resettlement responsibilities can be actioned. Some of the digital exclusion experienced by this population could also likely have been avoided through government functions and services remaining open and in-person with safety protocols in place, out of a

Figure 4.8: Responsibilities in bridge building

recognition that there are certain segments of populations for whom digital services will never be a good option.

To mitigate the administrative barriers and challenges they experienced, many refugees and asylum seekers turned to the organisations they know and trust. In turn, these organisations supporting refugees and asylum seekers inserted themselves into the responsibilities related to the digital delivery of refugee support services in a bid to absorb these digital responsibilities for the greater good. As the study shows, refugee-serving organisations absorbed these responsibilities by spending significant amounts of time supporting their clients in ways often beyond their existing mandate – while also battling with their own organisational challenges of funding, remote work and an uncertain future. Examples include helping refugees navigate Universal Credit over phone sessions, filling out paper forms online, helping clients access online forms (all often while battling language barriers) and seeking to continue outreach and intake with a significantly digitally disadvantaged population. These refugee-serving organisations are from the voluntary and third sector, in contrast to being community responses or self-organising community networks. Such organisations are, as this study highlights, struggling with organisational insecurities related to funding, policy status and legal responsibilities.

Some of the situations and challenges shared raise concerns explored in the previous section about organisations' ability to manage data privacy and security. It also builds on past research on UK civil society's ability to manage data, which found that civil society organisations face barriers of limited time, funds and expertise to make use of data (Easton-Calabria and Allen, 2015). Although there have been calls to explore how 'civil society can play a role in shaping the direction and application of technology in policy and development', it is likely in reality that many organisations are struggling to cope, let alone lead, in the so-called digital turn. This is in large part because such organisations do not have a clearly identified position within a digital responsibility framework. However, this is in fact hard to know as civil society organisations are rarely included in

Figure 4.9: A new framework

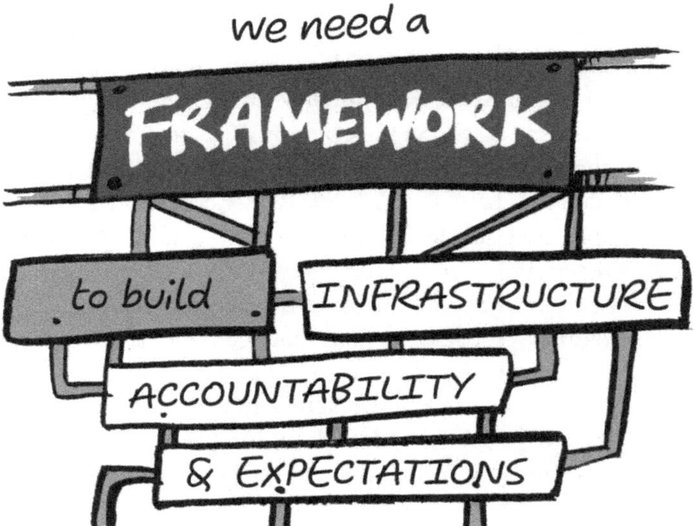

efforts to measure digital progress in society (Lynn et al, 2022). The findings show that such organisations find themselves in a position of trying to enable minoritised groups to action responsibilities realised and established through an outcomes-based framework. In a bid to become recognised as legitimate partners in this framework, refugee-serving voluntary and third sector organisations try to enable responsibility actions using the same outcomes-based framework. However, the findings suggest that a constitutively based framework might be more effective when working with refugee groups so that they can both build trust and negotiate responsibility actions to take different cultural practices and viewpoints into account.

In contrast to other sectors such as the private sector, where attempts have been made to create frameworks of digital responsibility for practice, no guiding framework of digital responsibility appears to exist for humanitarians and others engaged in refugee assistance. Some work posits that poor data management practices regarding refugees should be addressed through institutional capacity-building, such as including funding for data management systems and trainings into host government support (Schoemaker et al, 2018). At the same time suggestions such as these place the onus of digital responsibility in terms of data security onto host governments, other recommendations focus on the role that UNHCR plays in collecting identity data, such as the creation of a 'multi-stakeholder working group on interoperability chaired by UNHCR … to support a longer-term standards body focused on identity data' (Schoemaker et al, 2018, p 3).

The case study findings indicate that, aside from the digital responsibilities of personal data management, there is little or no recognition by the entities that realise and establish responsibilities for the resettlement and care of refugees that digital-first or digital-by-default policies create additional (digital) responsibilities that need to be inserted into the responsibility framework set out in law for the care of refugees. Instead, there is a greater focus on digital technologies being used to reduce the costs of refugee care and regulate migratory movement, thereby focusing on a digital responsibility towards indigenous rather than displaced or mobile populations. Digital responsibilities related to the care of refugee groups are, in contrast, almost totally silent.

Conclusion

Several compelling overlaps emerge from both case studies and the wider research, although organisations examined differed in size and scope. First, refugee-serving organisations seem to feel a responsibility to promote the digital inclusion of refugees, even if this is not explicitly within their mandate. Instead, this work towards digital inclusion comes in part through their efforts to provide other services, be it language classes or helping refugees secure jobs. In this way, whether in a context of resettlement or of global displacement, the responsibility to provide access to and literacy of digital technologies often comes out of a wider responsibility to provide support, rights and protection to refugees.

Among other points, this reinforces the centrality of technology to the everyday lives of some – though not all – refugees and asylum seekers and clarifies the importance of further exploring digital responsibility. At the same time, organisations often lack the skills and knowledge to digitally protect refugees the way they intend to, which represents an important area of increased attention related to both digital and 'in real life' responsibility. We would therefore argue that at a minimum, humanitarian agencies should issue guidance for staff to follow data protection and privacy standards outlined by the International Committee of the Red Cross (ICRC) Handbook on Data Protection, which provides detailed information on a range of digital technologies including mobile messaging apps, blockchain and AI to ensure that both risks and good practices are clear (2024). This literacy should also extend to refugees. As one key informant summarised,

> We need more participatory research and to show pathways where digital responsibility can really be implemented. This has to do with granting rights to refugees, such as partnering with actors to redesign pathways of asylum applications or digital asylum hearings. Refugees also need to understand that the digital identity they construct can affect the kind of opportunities and even the kinds of rights that are granted to them. Refugees need to understand how their data is being used. Humanitarian organisations – big and small – are responsible for organising data literacy and digital literacy practices towards refugees.

Second, some of this responsibility appears to come not only from organisations' own ambits or expertise but due to a lack of responsibility taken by others – host states lacking data and privacy protection regulations, the UK Government not fulfilling its own promises for increasing digital inclusion, and so on. How differently might have the work of some of these organisations been if the UK Government had provided smartphones as well as SIM cards to those asylum seekers that needed them, for example, or provided more direct support to civil society organisations working to address these barriers for refugees and asylum seekers themselves? The UK Government must further acknowledge and address the fact that its lack of responsibility-taking is in essence the delegation of responsibility to others with fewer resources to take it on. But responsibilities must not be delegated unless the delegator ensures that the delegated responsibilities are capable of actually being actioned.

From the findings and analysis, a variety of recommendations arise from this case study, which include:

- More government support (responsibility-taking) to help organisations fostering digital inclusion for refugees and asylum seekers in the UK.

- The development of a framework of digital responsibility for refugee-serving organisations that is relevant in international humanitarian settings as well as in domestic resettlement contexts.
- Increased sharing of information with refugees on how their data is used, on the level of online and other surveillance that may exist along migration routes and in countries such as those within the EU or in the UK.
- Further research could explore what the concept of 'refugee-centred digital responsibility' might look like for refugee-serving organisations and host and resettlement states, which places the needs, skills and interests of refugees at the core of digital inclusion, rights and protection today and in the future.

FIVE

The Digital Census and Kurdish Community Self-Organisation

Iida Käyhkö

Introduction

In this chapter digital responsibilities are examined through a local lens, considering how a digital deployment of a civic policy, in this case a population census, reveals how the state can take digital responsibilities for granted. It also shows how local actors are often required to step in and make it possible for individuals to action their civic responsibilities digitally. This is a study that can be framed as one situated at an intersection between digital responsibility and civic responsibility.

The study examines digital responsibility practices among London's Kurdish community in relation to the UK's 2021 census and provides input on creating accessible and secure digital services for marginalised communities, particularly those that have a fraught relationship with the state. It provides a window into the existing practices of responsibility among London's growing Kurdish diaspora, and what these can tell policy makers about the concerns and challenges encountered in digital interactions with the state. Further, it looks at how digital services are engaged within a particular cultural setting, and why specific cultural environments must be considered in discussions on digital responsibility.

Figure 5.1: Digital census

The study engaged with the largely Turkish-speaking North London Kurdish community, which settled in Hackney, Haringey and Enfield throughout the 1990s and 2000s. This is a multiply marginalised community experiencing housing and employment insecurities, the UK's hostile environment towards migrants, as well as criminalisation due to political activity. For this study, interviews were carried out with community organisers who designed and implemented a self-organised outreach and support plan for the 2021 census. Further, the social media output from Kurdish organisations and community members was analysed.

The study found that the idea of digital responsibility was shaped by specific cultural settings, and that practices of outreach and support may already exist in informal and semi-formal networks, unnoticed or unaided by local councils or central government. While the census of 2021 might have been expected to be met with suspicion among London's Kurdish community, steps taken by community organisers created a sense of trust and excitement around it and provided secure and accessible environments for community members to fill in the census. The shouldering of digital responsibility is not merely

Figure 5.2: The need for community response

a question of providing expertise or assisters to marginalised communities, but rather of engaging with existing community networks from a perspective of civic responsibility. By examining digital responsibility through a perspective of civic responsibility, this study urges a move towards an understanding of relational and communal challenges and resources in questions of digital responsibility, and the role of community self-organisation in providing both security and access.

The study begins with a brief outline of the context in which the research was carried out, focusing on the Kurdish community's relationship with the British state and with census-taking. Following this, the study design is outlined. This is followed by methods and findings, before a discussion section and conclusions.

Background

The 2021 UK census was a digital-by-default census or a so-called 'smart digital census'. This approach to gathering census data is regarded as having a number of advantages over a paper-based data collection approach. These advantages include (Younus et al, 2024): greater data accuracy, efficient and cost-effective data collection, reduction of data errors and a more comprehensive analysis of population trends and a more inclusive means of gathering data on these trends. Of particular interest for demographers are the trends in ethnicity. The 2021 census enabled further data gathering and analysis on the relationships between ethnicity and national identity in England and Wales. The design of the 2021 census included automated capacity to identity as ethnically Kurdish, and the use of a smart digital census enabled the first real-time data gathering on the demographics of the Kurdish diaspora in London.

The Kurds are the world's largest stateless people, with a significant diaspora population outside of Kurdistan. The area of Kurdistan is divided between four states – Turkey, Syria, Iraq and Iran – with the largest Kurdish population living within the borders of Turkey (McDowall, 2021). Kurds have been the target of genocidal and assimilationist policies for several decades across these four nation states. The UK is host to a significant diaspora community which is politically active in grassroots community organising and lobbying over the political rights of Kurds within Kurdistan (Jongerden and Akkaya, 2018).

As is frequently noted in introductions to the Kurdish people, no reliable census data on the Kurdish, or indeed Alevi, population exists almost anywhere in the world, including the region of Kurdistan (Sirkeci, 2000; van Bruinessen, 2000). The four nation states occupying Kurdistan are not keen to encourage Kurdish nationalism or political activity through providing opportunities to identify as Kurdish; further, forced assimilation efforts across the region have not only denied the existence of the Kurdish people, but also succeeded in erasing the ethnic and cultural background of many families and communities (Sirkeci, 2000).

In Europe, Kurds are largely included in Turkish, Iranian, Iraqi or Syrian citizenship and ethnic categories, reflecting the diversity in linguistic, national and cultural identifiers among Kurds. In the UK, the only way to identify as Kurdish in the national identity, ethnicity or language portions in the censuses prior to 2021 was through the writing-in option. In addition, the Home Office does not collate data on the ethnic background of asylum seekers, only their country of origin (Demir, 2012). Consequently, the precise number of Kurdish people in the UK is not known. The majority of the Kurdish population of the UK is subsumed into the 'Turkish' citizenship category – reflecting the Kurds' societal position as an 'invisible' minority population (Baser, 2011; Enneli et al, 2005). Kurdish community organisations emphasise that the lack of a straightforward way of identifying as Kurdish has led to many Kurds self-identifying according to the nationality they hold, rather than using the write-in option (van Wilgenburg, 2021).

Estimates from Kurdish community organisations regarding the number of Kurds living in the UK have ranged from 100,000 to 180,000 (Demir, 2012); these estimates, however, are not recent enough to reflect the arrival of significant numbers of Kurdish refugees from Syria following the civil war, nor those fleeing the escalation of anti-Kurdish political repression in Turkey in recent years. In the 2011 census in England and Wales, just under 50,000 people identified as Kurdish (ONS, 2011). A similar number of people stated their first language was Kurdish (Gopal and Matras, 2013). Both identifications were done through the write-in option.

This study was situated in the UK, with the Kurdish diaspora settled in London. This study built on an existing research relationship with the largely Turkish-speaking North London Kurdish community, which settled in Hackney, Haringey and Enfield throughout the 1990s and 2000s, with Dalston and Green Lanes as two local hubs of community centres and businesses (Wahlbeck, 1998). One of these community centres, in Haringey, is a central hub for Kurds across North London.

Much of the community actively involved in the day-to-day activities of this community centre considers itself part of the broad Kurdistan Freedom Movement, which is a network of broadly left-wing civil society organisations which has its origins in the PKK (Partîya Karkeren Kurdistan, Kurdistan Worker's Party). As the PKK is included in listings of terrorist organisations in the UK and in the European Union, the community experiences political repression and criminalisation in the form of arrests, raids and surveillance (Sentas, 2016). The British state's relationship with NATO ally Turkey is seen by many in the community as a security concern: many active community members are political refugees fleeing repression, whose extradition the Turkish state continues to demand (Sentas, 2018).

Further, London's Kurdish population experiences insecurity and criminalisation on two other levels. First, as asylum seekers, refugees and migrants under the UK's hostile environment policy, many among the community face bureaucratic challenges in accessing employment, housing and public services (Enneli et al, 2005). Racism and Islamophobia are common experiences, which further impact housing and employment opportunities as well as educational achievement (Jones, 2013). Second, Kurdish

Figure 5.3: Trust – a key challenge

men and boys frequently face discriminatory policing and stop-and-search practices which specifically target young racialised men (Ali and Champion, 2021). These experiences contribute to a generalised sense of suspicion towards the state.

Research approach

The study was designed to examine what we can learn about digital responsibility from looking at how a community with severe mistrust of the state engages with the state's data collection enterprise. The context for the study is the completion of the 2021 UK census that was designed as a digital-by-default activity. As part of the study, responsibilities for the following were considered: barriers to digital participation among London's Kurdish community, the role existing community networks play in facilitating engagement with the census, and community mitigation of anxieties concerning the use of personal and sensitive data when facilitating interaction between the state and individuals.

In April 2021, at the tail end of the third COVID-19 lockdown across England and Wales, the Office for National Statistics commenced its once-a-decade flagship project in a new format: the first census in which the majority of data was collected online. As the census moved online for 2021, the possibility for assisted, 'search-as-you-type' write-in options provided a greater number of officially recognised options for national identity, ethnicity, language and religion. The ONS ran a public consultation on tick box provision for ethnic groups, with certain modifications made for 2021, and stated: 'Write-in answers, online "search-as-you-type" functionality, together with provision of communications, community engagement, and advice and help from field force and contact centre staff, will ensure everyone can tell us how they wish to identify themselves' (ONS, 2020).

Kurdish was included for the first time as a 'search-as-you-type' option for national identity, ethnicity and language, while

Alevi was included for the first time as an option for religion. This had been lobbied for by a number of Kurdish and Alevi community organisations over the years and was celebrated widely as a crucial moment of recognition for the Kurdish diaspora community, with potential geopolitical repercussions (van Wilgenburg, 2021).

However, the COVID-19 pandemic hindered much of this work, as noted in a Parliamentary briefing paper: 'The COVID-19 pandemic poses operational challenges to census-taking, with the potential to affect both the preparation phase (e.g. engaging with community groups), and the census itself (e.g. carrying out face-to-face work to support people to complete returns)' (Barton and House of Commons Library, 2021).

Census information was provided to all households via post, with directions to responding online and an individual reference number included. Much of this correspondence emphasised the fines households might accrue by not responding to the census. To aid with non-digital census responses, field staff were sent to knock on doors across England and Wales. Many local services, libraries in particular, offered support in filling in the census (Barton and House of Commons Library, 2021).

Study design and methods

This study approached known community organisers with semi-structured interviews relating to their support and outreach efforts for the 2021 census, with the following interview questions:

- What were your concerns regarding the census of 2021 and the Kurdish community?
- What were the aims of the outreach and support work you helped carry out?
- What kinds of obstacles/concerns did people in the Kurdish community experience/express regarding the census? Were there positive aspects to it, too?

- What support did you receive for this outreach work from local government?
- How do you feel about state services increasingly moving to digital platforms? What impact do you think this will have on the Kurdish community in the UK?

Further, this study looked at social media content produced by members of the community to promote participation in the census.

Findings

The main findings focused on the role of community organisations in mediating the census requirements from the state and capability gaps and anxieties of members of the Kurdish diaspora. The first theme sets out the actions that the community organisations undertake to encourage and enable individuals to complete the census. The second theme sketches the main barriers to access for this particular group of census respondents. The third theme explores how the completion of the census keyed into the wider security concerns experienced by this participant group. The fourth and final theme reflects the disconnection between the establishment of the responsibility to complete the UK census, as set out in the instructions from the UK Government, and the avoidance of the responsibility – as perceived by the community organisers – by the local council to help individuals complete the census.

Support and outreach work

Community organisers at the Kurdish community centre in Haringey carried out an outreach and support project to help provide information on the census, as well as practical support in responding to the census. Outreach activities included:

- Creating social media content to be shared across different platforms, including on community WhatsApp groups.

Figure 5.4: Types of outreach

- Running adverts in local North London newspapers to publicise available support.
- Providing short clips to be run on Kurdish TV channels available in the UK.

These outreach messages emphasised the importance of filling out the census, included some practical visual guidance on responding to the census and highlighted that, for the first time, the census included 'Kurdish' as an option for national identity, ethnicity and language and 'Alevi' for religion as 'search-as-you-type' options. Some of the activities undertaken were modelled on outreach efforts for previous censuses, and the familiarity of community organisers with other digital support efforts in the past; however, they noted that the provision of these 'search-as-you-type' options added to their sense of needing to provide support for filling in the census.

In addition, community organisers assisted in filling in the census over a six-week period at a community centre in Haringey. Every Sunday, as well as on one weekday night, four community members equipped with laptops would assist, translate and in many cases fill in the census for community members requiring assistance. The community organisers were from a younger generation of the community, all proficient in English and confident with their technological skills, but with no formal training in offering such services. However, as is common among children of migrants, many among this generation of Kurdish community members have been supporting their parents or families in relation to asylum claims, citizenship accession and other legal and/or bureaucratic processes from a young age (Fisher et al, 2016). Further, these community organisers were all well known in the community through their involvement in community organising. There was an expectation from the Kurdish community that they would make themselves available to support with census responses; this occurred within existing organisational and interpersonal relationships of trust and accountability.

Figure 5.5: Targeted census support

Social media content produced by community organisers and members included videos of different lengths, as well as infographics featuring GIFs of box-ticking or writing in 'Kurdish' or 'Alevi'. Much of this content was circulated widely across Instagram, Facebook and Twitter by organisational and individual accounts. This often-humorous content shares certain elements: explaining the significance of the census for local issues such as funding, or other state support for Kurds and Kurdish organisations; and just as frequently, emphasising the importance of making the UK's Kurdish population count in questions of foreign policy. A video, produced by young members of the Kurdish community, which premiered on YouTube in March 2021, pokes gentle fun at Kurdish cultural stereotypes while providing accessible information about the census in Kurmancî Kurdish. In the video, a census field worker arrives at the door of a North London Kurdish household, is promptly invited in, with a lavish spread of tea and food provided. The video shows the census field worker asking questions about the inhabitants of the household, the work they do and their national identity, ethnicity, language and religion, with the mother stating that they are all Kurdish, speak Kurdish and belong to the Alevi faith. At the end of the video, the mother invites the census worker, at this point addressed as 'sister', to stay the night, in a portrait of Kurdish hospitality (Sinayic, 2021).

Inclusivity and access

Community organisers outlined three central concerns for the Kurdish and Turkish-speaking community in London regarding filling in the census: access to a computer, technological skills and English language proficiency. Interviewees also noted that many older members of the community are illiterate or not confident in their ability to read. While many people received some support from family members, often children of different ages, the support provided by the community centre was

Figure 5.6: Technology barriers to overcome

also key to being able to and confident in responding to the census. The community organisers emphasised that despite the outreach work, many of the community members who received assistance in responding to the census were visiting the centre for other reasons: language classes, cultural events or on social visits. "We would grab them as they came in", one interviewee put it. The social nature of these interactions was emphasised by all interviewees: they described having tea together with community members while filling in the census or having community members offer dried fruit or homemade treats in thanks for the help they received.

Community organisers all emphasised the sense of excitement generated by the census:

> [The census] created a huge excitement in the community, and a sense of belonging [because it] gave us the opportunity to declare who we are … I heard my mum's neighbours talking about it, I heard aunties and uncles at weddings and funerals asking each other, "Have you filled it in yet?" (Interview 1)

This was also the main focus of informational social media content created by the community: emphasising the chance to identify as Kurdish and/or Alevi. All interviewees noted the sense of acceptance and belonging engendered by the inclusion of 'search-as-you-type' options, with many community members saying that seeing 'Kurdish' and 'Alevi' come up in the census form made them feel recognised by the state, and therefore more a part of British society. This was felt to be simultaneously a gesture of recognition and goodwill on behalf of the British state, and a demonstration of successful lobbying efforts by Kurdish and Alevi organisations:

> [The census] gave a feeling of being part of the wider British community. A Christian country giving us the chance to declare who we are! People really feel that this was a chance to show the British state that we are here. (Interview 2)

> I was excited to be part of the census. It was an opportunity to say, "I am a part of this community." (Interview 3)

Insecurities and concerns

On the other hand, community organisers noted that some members of the community were afraid of the possible consequences of identifying as Kurdish, Alevi or both: "Two or three families I helped [respond to the census] had this worry of declaring themselves Kurdish and Alevi. They were asking me, 'Will this cause problems when we travel to Turkey?' People are so worried about imprisonment and harassment" (Interview 1).

Specifically, this fear was related to the sharing of personal data with the Turkish state, and the possible negative consequences that might be encountered when travelling to Turkey to visit family, or consequences for family members living in Turkey. Interviewees pointed out that as political migrants, many community members have personal experience of persecution and repression due to their ethnicity or religion. Community members generally were not trusting, or feel inadequately informed, about who their personal information would be shared with. It is significant, however, that none of the community organisers spoke directly about questions of security, indicating that these were not a prominent element in their interactions with community members. When asked, they did note these fears over data sharing with the Turkish state. However, they were not able to recall specific phrases or facts they might have used to assuage these fears. Although the ONS census website included information on data confidentiality, this was not mentioned by any of the community organisers as having been a useful tool for them in communicating with community members. The provision of simple reassurance was enough. Similarly, most of the social media content shared by community organisers did not seek to explain the privacy and security of personal data provided in the census, nor the possible repercussions of not filling in the census, but maintained a focus on the more positive aspects of the census.

Many also felt a lack of trust in the British state and a lack of expectation of future inclusivity:

> This is just the government collecting information for themselves. Our expectations of the British government are not matched. Just a census isn't enough. (Interview 2)

> A lot of people were asking, "What are you doing for Kurds now?" We had this option [of identifying as Kurdish/Alevi] on the census, but we still only have a write-in option on council forms. Nothing has been provided for the community. Why isn't the government

providing a GCSE in Kurdish, or including Alevism on the curriculum? (Interview 1)

Sharing of responsibility

This feeling of the census not being enough to mitigate lack of support for the Kurdish community was felt particularly by the community organisers interviewed; they had received no support from local councils in their outreach and support work. When asked about their expectations regarding the council, the interviewees varied in their responses: from expressing no expectations of help, to articulating disappointment – particularly that local councillors from a Kurdish and/or Alevi background were not at all involved. None of them, however, expressed a feeling of surprise over the lack of support. It is important to note that it is possible that other Kurdish and/or Alevi community organisations in London received support from local authorities, but that there was no such outreach for this particular community centre.

Interviewees noted that without greater resources for outreach and support work, they felt they ultimately only reached a limited part of London's Kurdish community. They also emphasised future issues related to the British state's relationship with the Kurdish diaspora community; one interviewee pointed out that the struggle to carry out adequate outreach and support work for digital service access will continue for at least the next two or three decades, "until everyone in the community is digitally capable".

Finally, interviewees emphasised that in general, trust towards the state is minimal among much of the Kurdish community. They noted that many community members have been living and working in the UK for decades and have faced harassment by the police throughout. One interviewee characterised the British state as making few efforts to create bonds and warned: "In time, it will be impossible to come together."

Thinking about digital responsibility

Three lessons on digital responsibility can be learnt from this case study of the census outreach and support work carried out by the Kurdish community in North London:

1. Inclusivity is key to accessibility.
2. Trust and a sense of security are important in mitigating security concerns.
3. Community self-organisation can play an important yet invisible role in supported access.

First, all community organisers emphasised the significance of the recognition of Kurdish and Alevi identities in the 2021 census, and most of their outreach output emphasised this aspect of the census over all others. While it may have been expected that there would be hostility towards the census among a community that has a fraught relationship with the British state, this was not the case: the excitement and sense of belonging generated by the inclusivity of the census design far outstripped any concerns. There is a growing body of literature within digital civics, human computer interaction and critical security studies that focuses on describing difficulties of access among marginalised communities, and the many governmental, government-supported or grassroots efforts to facilitate access (Jensen et al, 2020a, 2020b; Talhouk et al, 2016). A growing body of literature emphasises the need for involving migrant communities in the design process to provide communities with secure and accessible digital services (Almohamed and Vyas, 2016; Fisher and Yafi, 2018).

The importance of inclusivity – even small acts such as 'search-as-you-type' options – may generate an unexpectedly large sense of goodwill and even community mobilisation efforts. Significantly, the number of people identifying as ethnically Kurdish in the 2021 census almost doubled from the 2011 census, from slightly under 50,000 to over 90,000.

Speculatively, one might surmise that the provision of the 'search-as-you-type' option, alongside community outreach both in person and online, together had some impact on this increase in numbers. Importantly, as this study shows, simply making the census digitally available did not make the census accessible. There was a clear disconnection between the realising and establishment of the responsibility of individuals and households to complete the census (as set out by the UK's central government) and the ability of individuals to respond and carry out this responsibility. In order for actioning of the individual's responsibility to take place, community organisations had to realise, establish and absorb the responsibility of making the census accessible to individuals. The findings also show that the excitement and sense of belonging generated by the inclusivity of the census design created the conditions under which community organisations were willing to insert themselves in this way. More must therefore be done to increase inclusivity across different spheres of digital services.

Second, this study highlights the significant role played by trust and a sense of security in responding to security concerns and underscores the significance of the relationship between responsibility and security. This relationship is no less important in the digital context and where the responsibility framework

Figure 5.7: Digital innovation needs inclusive responsibilities

is outcomes-based the likelihood is that digital protections and legal compliance will be the security approach, and where the responsibility framework is constitutively based the likelihood is that relational support, consensus building and trust-building will be the security approach. While none of the community organisers were able to – or even aimed to – provide detailed information to community members on the security of personal data in the census, their trustworthiness and accountability, and the safe, familiar space they created for the outreach work, played an important role in allaying security concerns. The role of relational services and access provided by kin or community networks on which many marginalised users depend is frequently characterised by networks of trust, which often operate very differently from more formal supported access efforts (Chouhan et al, 2019; Cipolla and Manzini, 2009; Ruiu and Ragnedda, 2016; Selwyn et al, 2005).

This relational approach foregrounds the role of ontological security, of feeling a sense of security, in accessing digital services (Giddens, 1991; Roe, 2008). As noted by Coles-Kemp and Jensen (2019, p 181), it is 'an individual's senses of trust and confidence that enables and sustains effective access' Marginalised people face heightened material and ontological insecurities in interacting with the state, often relating to asylum or residency claims (Jensen et al, 2020a). This study shows that it is not merely the provision of information that is central to reducing a sense of insecurity concerning digital services, but also the relationships, spaces and existing sense of trust that communities can offer. These relational services in turn provided the pathways through which responsibilities are actioned and in so doing, fostered a sense of positive security and enablement that reduced security anxieties. As the findings show, the choice of action could be tailored through the interactions between the community organisations and the individuals. The relational services also helped to establish a credibility of the civic and digital responsibilities that had been established around the delivery of the census.

Figure 5.8: Dispersed responsibilities

Finally, this study shows the significance of community self-organisation – often unnoticed or unaided by local councils or central government – in providing access and a sense of security. The shouldering of digital responsibility as carried out by the Kurdish community organisers interviewed in this study does not relate to the frequently evoked themes of *blame* and *culpability* for designers, innovators and organisations within digital responsibility literature. For example, van de Poel and Sand (2018) discuss the problems of attributing responsibility in innovation, proposing that accountability processes must be able to confirm the 'moral authorship' of innovators, and that a virtue-responsibility approach should enable innovators to take on responsibilities actively and voluntarily. This approach – of having to *create* a sense of responsibility for when things go wrong, and to coerce, or preferably to convince actors to take it on – is an underlying theme across much of the digital responsibility literature (Lima and Cha, 2020; Markus and Mentzer, 2014). It does not, however, describe the central questions of digital responsibility explored in this study. Instead, the community organisations adopted a constitutive form of responsibility, which is forward-looking and largely uses social practices to negotiate and carry out responsibilities.

The study also provides an alternative framing of *responsibilisation* – which is typically regarded as governments requiring individuals to take action to reduce their vulnerability to cybersecurity issues, or to access digital services (Renaud et al, 2018). In this instantiation of responsibilisation, Kurdish community organisers independently took on the responsibility of providing digital access to the census for community members: community organisers did not expect support from local councils or central government, but did express their disappointment at the lack of it. Specifically, they articulated a wish for help in their outreach efforts and publicising the support work they were offering. Importantly, they were happy to self-organise support. The move to insert themselves between the state and the individual shows that community organisations understand that you cannot have effective responsibilisation without ensuring that the individuals can

Figure 5.9: Real relationships and digital responsibilities

action the responsibilities that are being conferred upon them. The provision of support by trusted members of the community in a shared cultural context and familiar space contributed to the access and sense of security felt by community members in relation to filling in the census. Therefore, it is crucial for local councils and central government to appreciate and understand existing relationships and community efforts in the realm of digital responsibility, and to support these on the terms of communities in question, especially about the agreement of all parties to monitor and regulate across the digital environment.

Conclusion

This study shines a light on the complexities of actioning digital responsibilities and highlights the disconnect that often exists between the civic responsibilities realised and established by governments and the realities of actioning those responsibilities by digital means. This study also reflects how responsibilities might be realised and established using an outcome-focused responsibility framework but are actioned using a constitutively based responsibility framework. The study also shows how through actioning digital responsibilities, responsibilities are often re-distributed and re-negotiated in order to enable individuals to make the required responses.

The problems of access and security faced by London's Kurdish community are well understood by community members and organisers. The well-coordinated outreach and support efforts relating to the 2021 census reflect this. These problems range from ones commonly found among marginalised and migrant communities, to highly specific issues encountered by the Kurdish community. This study shows that the shouldering of digital responsibility is not merely a question of providing expertise or assisters to marginalised communities, but rather of engaging with existing community networks of trust and duty. By examining digital responsibility through a perspective of civic responsibility, this study urges a move towards an understanding of relational and communal challenges and resources in questions of digital responsibility, and the role of community self-organisation in shouldering a range of responsibilities regarding security and access. This study also indicates the need for further research on the differences and overlaps of formal and informal support, as well as the specific needs of different communities in accessing digital services.

Digital responsibility efforts should seek to build on already existing, everyday relationships between people and organisations to create shared parameters and practices of access, security, inclusion and responsibility. Leaving informal or semi-formal networks and communities already shouldering digital responsibilities unsupported and unacknowledged creates a sense of disconnection and resentment among communities, and disregards opportunities of developing shared responsibilities. A greater focus on inclusivity and collaboration with communities would increase the accessibility and security of digital services in particular for marginalised communities. Doing this design, outreach and support work on the terms of communities and with respect to existing networks and spaces of trust would be more likely to create a positive feedback loop of inclusion and shared responsibility. Supporting self-organisation efforts is a chance to create dignified, culturally sensitive digital inclusion in the future.

PART III

Understanding Digital Responsibilities

SIX

Lessons Learned

Introduction

We now bring the book to its substantive conclusion by considering what the previous chapters tell us about digital responsibilities and draw lessons that can be learned relevant to future digital policy making when our framework is applied. Applying the framework, the case studies can be understood as digital policy challenges in different ways:

- Chapter 3, Digital Safety, is a principal challenge of *realising responsibility* and specifically when responsibility is not realised, the state falls back on ascription of blame through the establishment of legal obligation and technological enforcement to achieve responsibility actions.
- Chapter 4, Digital Migration, is a challenge of *establishing responsibility* which does not create sufficient impetus for positive responsibility actions which then leaves vulnerable refugee communities looking to non-state for protection and digital support.
- Chapter 5, Digital Census, is a challenge of *actioning responsibility* to absorb the limited discharge of realised and established responsibilities by central and local government that created a lacuna which required civil society institutions and organisers to step in.

Five lessons can be adduced. First, even though all three case studies highlight responsibility actions, we argue that actions do not tell the full story about digital responsibilities. The actions are fragmented and must be understood in the context of realising and establishing responsibility. It is that context which helps to better situate and explain the actioning strategies across all case studies. Second, all case studies highlight the distributed nature of digital responsibilities including the distribution of responsibilities across infrastructural networks of digital technology that are governed by different legal and regulatory requirements. Responsibility distribution, then, is a key facet of each case study but also a key determinant factor in our focus on realising, establishing and actioning.

Third, the distributive nature of digital responsibilities also leads to lacunas across realising, establishing and actioning processes. Lacunas are absent spaces, devoid of responsibility allocation, which require actions to be taken by parties for digital technology implementation to take place. The lacunas that arise in the case studies are mostly at the actioning level where the positive actions of some parties obviate the attempts to avoid or refuse responsibilities by others. Responsibility lacunas are consequently important because they highlight the limits of both legal obligation and security technologies. Fourth, as the case studies show, the tried and tested response to legal and technical lacunas is force, either through legitimate state action in the establishment of legislative obligation or the use of security controls in digital technology. The case studies highlight the limits of legal and security forcing functions that essentially seek to bypass collaboration and consensus to achieve a responsibilisation outcome. Finally, we then consider the role that resource allocation by government and revenue generation by the private sector played in the case studies.

All of this leads to where we started our book with an overview of different responsibility frameworks. We contend that responsibility frameworks need to be understood differently, regardless of disciplinary orientation. Largely speaking, the traditional frameworks of responsibility are grounded in an older model of individuality that does not fully account for the multiplicities that arise from the digital context. As we show in our lessons learned discussion, ascription of legal rules and security controls will only go so far and therefore responsibility frameworks must be broader in their thinking about the need for consensus and collaboration. With that in mind, let's consider in more depth what lessons can be learned, beginning with what the case studies tell us about responsibility actioning and why it is important to understand actions in the context of realising and establishing responsibility.

Actions tell us a lot about responsibility (but not everything)

Chapter 2 highlighted different responsibility frameworks from different literature. It outlined that responsibility is an amorphous concept with tendrils that run deep into moral philosophy, social and political science, technology development and law. The tendrils run so deep that responsibility considerations tend to surface through actions. It is unsurprising then that responsibility frameworks are action-based, whether they be the blameworthy ascriptions of traditional legal approaches or their contemporary applications in responsibility gap analyses. Even contemporary frameworks, such as RRI, are grounded in the broader governance of responsible acts around technology development. Actions are thus an important way to conceptualise and understand responsibility. The actioning of responsibility also has the advantage for scholars of being visible which means it can be determined and measured to ascertain whether an individual or party is responsible for their acts or accountable with their obligations.

Actioning has an important role in our framework and we agree that a starting focus on responsibility actions is informative. Take, for example, the application of the case studies to the non-exhaustive list of responsibility actions that we proposed in Chapter 2. Doing so enables the seeing of

Figure 6.1: Responsibility actions

responsibility in a clearer way that specifically acknowledges the possibility of both positive and negative acts.

1. *Absorb* – the highest positive act as it seeks to fulfil the agreed consensus of the environment at potentially the expense of the individual party's own benefit. Several acts of absorption are identifiable from the case studies, particularly relating to the Digital Migration and Digital Census chapters. Civil society organisations in both cases voluntarily absorbed the established responsibility of other parties to provide extra support for vulnerable communities. Chapter 4 highlighted the actions of underfunded refugee-serving organisations and their attempts to extend the use of their own technologies to individuals and refugee communities. COVID-19 lockdowns exacerbated digital disparities and further marginalised refugees who were already struggling with the costs of internet access, device ownership and English language digital literacy requirements. In many ways, as highlighted in the chapter, the need for digital inclusion strategies for refugees was realised and established in international governance. Yet it had to be absorbed by community groups even in a broader society that was increasingly presented by government as 'digital-by-default'.

Chapter 5 also highlighted the extraordinary acts of care enacted through the support and outreach work of Kurdish community groups. The absorption of responsibility by those groups extended from the training required to successfully operate a rather simple digital solution to creating a shared cultural experience that assisted to allay the security fears of the Kurdish community. Absorption then is double fronted. First, there is the willingness to be responsible for the implementation of a digital measure that is essentially beyond the party's remit, but still within its capabilities. Second, and more importantly, there is the creation of a culturally safe environment to generate support and interest on the basis that it will carry longer-term benefits for the community. Both absorbing acts are way beyond the implementation strategies envisaged by the state.

2. *Allocate* – in this situation, parties take on responsibilities designated to them, and to others, to identify and share those responsibilities across environments. In many ways, responsibility allocation is a classic act of government and a key method for allocating responsibility actions is by the implementation of law. Digital Safety's coverage of the pre-legislative debates leading up to the OSA is a clear example. The allocation of new duty-based responsibilities to online service providers outlines the allocative role that traditionally government has played. Allocation, in this sense, was not based on the consensus requirements we called for in Chapter 2 and was ultimately forced upon perceived recalcitrant parties who were unlikely to bear responsibility for actions but were nonetheless best placed to implement. Equally, the co-ordination of community response, and the sharing of resources in relation to positive responses, was a key similarity in the Kurdish and civil society actions outlined in Chapters 4 and 5.
3. *Discharge* – refers to the baseline act of fulfilling or discharging a responsibility. The act of discharge sits in the middle of

the spectrum because it can represent both positive and negative acts. A discharge is thus contextual. An example of discharging behaviours arising from the case studies involves minimum level implementation strategies for refugee and asylum seeker digital services provided by local authorities. At times, these were bare minimum requirements in which it is questionable whether the established responsibilities were actually discharged. The same perhaps could be said about the provision of additional language capabilities in the census. The provision of a digital solution without the necessary support, training and community assurance required for implementation by the Kurdish community, could be viewed as a baseline discharge by the ONS.

4. *Avoid* – refers to the negative act of avoiding a responsibility that has been realised and established as relevant to a party. Digital Safety highlighted that many of the contentions arising from the OSA's fractious debates regarded online service provider attempts to avoid responsibility for the broader protection of digital environments. Consistent attempts to reframe problem construction as a matter of content moderation, in which individuals were better placed to respond, can be seen as an avoidance of engagement with the problems highlighted by all the other parties within the broader digital environment. Similarly, the actions of local authorities, and especially of central government, could be considered avoidance of responsibility in relation to the provision of digital services for refugees in Digital Migration.
5. *Refuse* – a definitively negative act of refusing to take on a responsibility that has been realised, established and actioned by other parties flowing from collaboration and consensus. The case studies show that the impact of avoidance and refusal acts means that realised and established responsibilities eventually emanate on individuals through responsibilisation strategies. As such, even though none of the case studies specifically involved a refusal action, the extent that different parties tried to avoid responsibility had

> the same sort of impact upon individual citizens, namely, to try to make them responsible for someone else's established actions. The attempts to make users responsible for content moderation by online service providers and the inaction of local authorities in both the Digital Migration and Digital Census case studies, led to the need for absorption and reallocation of responsibility actions.

Our focus on the actions arising from the case studies helps to better understand the dynamic nature of responsibility. An action in one context can give rise to the need for a different type of action in another. An avoidance of responsibility by one party gives rise to a corresponding need for absorption or reallocation by another. Responsibility then is diffuse, shifting and amorphous. It changes as the actors and parties in given environments respond differently over time. It is not surprising therefore that the varied academic attempts to understand responsibility begin with actions. However, as alluded to already, we contend that it is equally important to understand that responsibility does not end with actions. Focusing on responsibility as actioning does not conceptualise the true complexity of responsibility, especially in the digital realm. Actions alone do not tell the full story. They are fragments of realising and establishing responsibility and so it is important to take a step back and consider digital responsibilities beyond actions. Actions are a sensible and useful starting point but they do not explain the complete picture.

That is why our framework also gives weight and prominence to realising and establishing responsibilities. As the case studies show, how responsibility is realised and established, or not as in some case studies, can impact upon how responsibilities are actioned. Understanding responsibility as purely a process of ascriptive output does not encapsulate the broader collaboration and consensus requirements that realising and establishing processes speak to. Without those processes, responsibility allocation merely returns to the ascription of individual blameworthiness and the historicised projection of future obligation. All of which encourages an understanding of responsibility that is predicated on the use of forceful authority to ascribe obligation upon individuals whose ultimate role is to tacitly accept. We now contend that traditional model is made more challenging because of the distributive nature of responsibility in digital environments.

Distribution of responsibilities is inescapable

Our framework's focus on realising, establishing and actioning also outlines the distributed nature of digital responsibility formation. The case studies highlight that the formation of responsibility does not take place at one specific site and is formed through different processes. There is a clear relationship between the top-down focus of established policy positions and ground-up actions across all three cases. For example, in the Digital Census case study, the originating act of realisation began when the ONS employed a drop-down list for different languages in the first digitally focused census. The use of the drop-down list would have been based on an established legal justification and actions were taken on the ground by Kurdish community groups to ensure its successful implementation across the diaspora. Similarly, the Digital Migration case study highlights that responsibilities were realised and established internationally which required actioning by central government and local authorities. As noted in the chapter, the actual actioning of responsibility, like the Digital Census chapter, was undertaken by civil society organisations. Equally, the Digital Safety case study highlighted the shifting movements of responsibility imposition through legislative development.

Traditional legal models of responsibility are based on a rather unitary understanding of how responsibility is formed. Law is created, obligations are formed and individuals accept responsibility for legally bounded actions as a recognised

Figure 6.2: Distributed responsibilities

social good for all. In one sense, this traditional model is still partly relevant to understanding the case studies through our framework. There are clear acts of forcible ascription, most notably in the Digital Safety case study. However, ascription alone does not tell the full responsibility story. The state's ability to ascribe responsibility through legislative obligation is still, of course, a prime source of institutional authority. That has not changed, and we are not contending that it has. However, what has changed are the distributive capabilities of digital technology which now enable responsibilities to emerge through technological infrastructure. It is the infrastructure that allows responsibility to be formed and actioned across many different environments.

As we consider further, the pathways of digital responsibility formation are consequently pluralistic. Take the Digital Safety case study as an example. The formation of the OSA highlights the difficulties involved in establishing digital responsibilities. In one sense, it could also be argued that the realisation of responsibility was never fully accomplished because of the shifting nature of legal establishment. There is already a movement, or distribution of responsibility realisation, that shifts across different parties at contrasting times. The Bill's life germinated from the Online Harms White Paper which had a multitude of harms that reached from national security to individual harms. The harms were so broad that necessary responsibility conversations were never fully realised, which helps to explain some of the difficulties arising for the Bill's establishment. Moreover, the Bill also entailed a confusing mix of command and hybrid governance that obliged online service providers to act but also allowed providers significant leeway to implement enforced duties. There was both an imposition of control through ascription and forms of delegation deemed necessary for a digitally distributed environment.

The different regulatory strategies at play highlight another factor pertinent to distribution considerations. The legal models of ascription, propounded by Hart and others, have traditionally guided how responsibility in a legal sense should be understood. As noted already, it was understood in a unitary sense which was representative of the forms of centralised institutional authority that characterised public policy and service delivery in the last century. We noted in Chapter 2 the effect that privatisation strategies had on understandings of responsibility formation in the context of responsibilisation, namely, by making citizens responsible for their choices and part-operation of services. The effect of privatisation, which Hart could not have contemplated, ranged from the decaying of the dominant public institutions that form responsibilities (Parliament, agencies, regulators, etc.) to the process of shifting responsibilities to the private sector and individuals. All these developments were silently predicated on a functional distribution model that now attached a range of responsibilities to different parties through digital infrastructure. New digital opportunities – digital census, monitored regulation of digital spaces, meeting the digital needs of the most vulnerable – equally came with a set of responsibilities for implementation that were now dispersed over location and mediated through infrastructure.

Contemporary responsibility frameworks, such as RRI, better contemplate the possibilities of distribution but they nonetheless do so from a confined perspective, namely, the responsible implementation of a given technology innovation, by a given set of actors. Even digital civics, with its much broader critique of public service delivery, does not consider the full consequences of responsibility distribution. We contend that our framework starts to capture these complexities. The case studies highlight how responsibility is distributed across different sites and realms through the interaction of the market as a government service delivery mechanism, including the absence of that mechanism and the impact on users, both as consumers and citizens.

The case studies also highlight the distribution of responsibility through the different processes of realisation, establishment and action. Movement of responsibility formation is fluid between these processes and is distributed dynamically. Finally, the case studies also show how responsibility is distributed across time, as the collaborative and consensus building elements of realising, establishing and actioning play out across different temporal frames. When all these aspects are combined, they demonstrate that responsibility formation is distributed, which requires a much stronger focus on collaboration and consensus processes.

The importance of lacunas

Chapter 2 highlighted a contested academic discussion about the existence and utility of responsibility gaps. To recap, there are two broad positions. The first, as espoused by Matthias and others, is that there is the potential for gaps of responsibility ascription in autonomous systems because there is no human decision maker that machine-based decisions can be ascribed to. In other words, the system has become so sophisticated that decisions are constructed in such a complex manner that they cannot be parsed to identify the causal relationship between output and human data inputs. It therefore becomes impossible

Figure 6.3: Responsibility lacunas

to ascribe responsibility to an individual in these circumstances. The 'responsibility gap' argument is not universally supported. Critics outline that even in complex automated systems it is still possible to ensure that a human is always made responsible for the outcomes of a machine-based output. The responsibility gap can be filled and responsibility can always be attributed. Responsibility is therefore constructed to meet circumstances rather than an accepted consequence of machine intelligence.

There are clearly some distinctions between the book's case studies and the type of digital infrastructure envisaged as part of the responsibility gap analysis. The case studies do not cover or examine the prospect of autonomous machines. In that sense, the digital component to our case studies is relatively low level, such as the use of online drop-down forms or basic digital support needs. The contentions involving digital safety do, however, speak to the socio-techno visions of autonomous machines. However, the focus of contention is not about the prospect of autonomous machine activity; rather, it is about the business models of online service providers and the appropriate role of the state in regulating those models. The business models are, of course, built to combat scale and are largely automated in their data collection and analysis processes. Nevertheless, they are not of the complexity contemplated in the gap analyses. We therefore do not use our book to contribute to the ongoing and broader debate about whether responsibility gaps do exist, and if so, what they tell us about responsibility in a digital context. However, that does not mean to say that the concept of gaps has no relevance to our overall argument.

We believe our case studies highlight responsibility lacunas rather than gaps. A lacuna is preferred because it speaks to the idea of an absent part that thus draws attention to something, rather than a gap that is required to be filled. It is the drawing of attention that becomes the important aspect of our analysis. We do not believe that responsibility lacunas are bugs of the system caused by the inability to assign responsibility to a given machine output. To follow the well-trodden logic, lacunas are a feature, rather than a bug, in the formation of digital responsibilities.

Disruptions caused by digital technology and infrastructural implementation will always cause lacunas. In part, this is due to the disruptive nature of novel technological application which will give rise to unexpected uses and unforeseen consequences. It is also due to the distributive nature of digital responsibility allocation as already outlined. Put simply, distributed responsibilities advance the prospects of lacunas emerging because there are going to be points of digital technology implementation where it is unclear how the technology will apply and who will be responsible for actioning. We consequently believe that a digital responsibilities framework must operate within this reality and accept that the dynamic nature of responsibility as realised, established and actioned means that formation will not always be readily visible. As noted in the previous section, actions can be seen as fragments of realisation and establishment and are thus, by nature, incomplete.

Lacunas are important because they highlight responsibility contentions that need collaborative processes to form consensus. Lacunas, in this regard, are a positive aspect of responsibility discourse that help to realise and establish responsibility. Take, for example, the lacunas arising from the case studies. Digital Census details the absence of an educational support process to augment the positive implementation of language drop-down lists. Seeing this absence purely as a responsibility gap does not fully encapsulate the uptick of positive actioning that flows from the Kurdish community. The absence of appropriate support mechanisms was not considered by that community as a space where no responsibilities exist. Instead, it was viewed as an opportunity to extend existing community bonds and to contribute to broader policy goals through census data provision that would lead to greater recognition both locally and nationally. Much like the types of non-exhaustive responsibility actions that we highlight in our framework, lacunas can have both positive and negative effects. In the Digital Census, and indeed the Digital Migration case study, the recognition of a lacuna led to positive actions. Even in the Digital Safety case study, the lacuna arising from the attempted avoidance of responsibility by online service providers also gave impetus to legislators to act positively.

We believe the identification and analysis of lacunas is important in understanding the formation of digital responsibilities. These are not spaces that should be feared by policy makers seeking to implement policy or service delivery outcomes through digital means, particularly involving third parties. In fact, we believe lacunas have the opposite effect. They highlight spaces of potential contention, spaces of unwanted responsibility by some parties that do not by default have to be filled by responsibilisation strategies that automatically bestow responsibility on individuals. Digital Census clearly shows the willingness of community activists to absorb responsibility for educational support where little to none was provided alongside the implementation of digital innovation, albeit straightforward. The case study also shows the distributed nature of digital responsibility formation which takes place across different physical and digital spaces. Formation took place in the code behind a website and the various public and private sites in North London.

Lacunas are an important part of understanding the digital landscape, and in particular, the connection points between the digital and physical in which responsibilities are actioned. These are also the places that show the limits of digital technology reach. Lacunas can provide valuable insights for policy implementers about how responsibilities are formed across potentially diffuse parties with diverse needs and goals. We contend then that lacunas give rise to the possibility of dialogue that promotes consensus and may therefore reduce the need for forceful actions, whether through technology or legal means. We now consider that point further in the context of forcing functions.

The limits of forcing functions

An under-discussed and under-researched area is the interaction between digital responsibilities and security technologies. In digital-first or digital-by-default societies, there is a reliance on security technologies to not only protect data, services and technology but to also implement or operationalise public and organisational policies. For example, consider welfare delivered as digital by default. Setting up an account on the digital welfare platform and using that account to access digital welfare services, make welfare claims and provide evidence of employment seeking, rely on technological security mechanisms. They can include: identification and authentication processes to verify that the claimant is who they claim to be; behavioural pattern monitoring that checks the verification of the claimant and validity checks against financial and open-source data to assess the likelihood of additional income. Traditional security mechanisms include identification and authentication, to ensure that authorised individuals can access services, data protection mechanisms to ensure confidentiality of data, and data integrity mechanisms, to ensure that data is not subject to unauthorised modifications. Even though these security functions can be extensive, as in the case of welfare monitoring, we nonetheless believe they are still limited in the context of digital responsibilities formation.

In addition to these traditional mechanisms, we have seen the growth of behavioural security mechanisms that use a combination of machine learning and AI to assess the authenticity of service user actions, to verify the assertions and claims that an individual makes about their circumstances and to assess the risk that the individual user is a threat to the system or service. Cloud computing and the development of sensor technology has also meant that these security mechanisms can draw on geographical, motion and biometric data to augment security control decisions. This contemporary paradigm of technological security controls has encroached on many aspects of our lives that were once thought of as separate and private and therefore require a greater degree of both trust and consent to work. Without trust and consent, such security technologies become a type of forcing function to push through the goals of public and organisational policy. When security technologies are used as a forcing function, the likelihood of workarounds, control subversion and non-engagement increases and with it the effectiveness of the security technology decreases.

Realising, establishing and actioning digital responsibilities in a way that is transparent and that creates consensus is an important means of increasing trust in state actors and

Figure 6.4: The limits of forcing functions

strengthening the effectiveness of security technologies. Reviewing our case studies, we can see that for security technologies to be used effectively to protect digital activity, and to enable individuals to safely access services, there needs to be consensus about where responsibilities lie and with whom. If consensus is not achieved, security technologies become less powerful because the duties and obligations necessary for them to be successfully deployed are absent. For example, in the case of age verification technology in Digital Safety, if no party recognises an obligation to ensure that such technologies are usable and accessible by all, individuals will be locked out of services and the technology will not be able to support the access control goal. Equally, if no party acknowledges that it has an obligation to respond to queries and potential errors in age verification, then the technology will not be able to support the access control. If no party ensures that age verification technology is deployed in a meaningful way, people will move to other, less controlled and potentially less visible, platforms leaving children more, rather than less, vulnerable to harmful content. Therefore, where there are lacunas, ambiguities and tensions in digital responsibility these should be proactively examined because such tensions can constrain the effectiveness of security controls and could thus inhibit successful digital policy implementation.

In some ways, the same could be said for the legal considerations outlined in the case studies. The pre-legislative development of the OSA highlights the difficulties involved in establishing digital responsibilities. In one sense, it could also be argued that the realisation of responsibility was never fully accomplished because of the shifting nature of legal establishment. The Bill's life germinated from the Online Harms White Paper which had a multitude of harms. From there, several tensions were perhaps never fully realised which helps to explain some of the difficulties arising for the Bill's establishment of responsibilities – for example, the transition of content harm to digital environmental protection that would have imposed new system-based positive duties on online service provider business models.

The development of the OSA, and indeed the other case studies, shows that even a strongly established piece of law does not necessarily lead to the successful implementation of the desired legislative outcome. This, of course, is well trodden ground in the public policy literature but we highlight it again now to outline the complex relationship between the attribution of state authority through the commanding of legislative obligation and the distributed nature of digital technology implementation. It is the uncertain combination of command and hybrid governance that can give rise to the type of responsibility lacuna we have already noted. In Digital Safety, it is the state that governs the shape and resolution of the absent part – to our mind the lack of consensus across all parties – with forceful authority. We acknowledge that this type of action is a constituent part of state authority. We are not defenders of the significant power that online service providers have amassed for themselves. Rather, we highlight this issue to acknowledge that use of law as a forcing function of legitimate state authority has limits due to the distributed nature of digital policy implementation. A better way of understanding the multifaceted contours of digital responsibilities is therefore required.

We believe that mapping digital responsibilities to show where there is discord as to where and how digital responsibilities should be ascribed and carried out is fundamental to the security analysis of a digital product and its use. Our case studies indicate that digital responsibility is defined at the intersection of: (i) the site where digital responsibilities are placed and (ii) the realm of activities that is underpinned by these responsibilities. The sites of digital responsibility (the state, the regulator, the market or the individual) are where obligations and duties are assigned. The environment's digital responsibility – information; content production and management; and design, management and implementation of technology – are where digital responsibilities are carried out. Our case studies show that there is a tension between the state, the regulator and the technology companies (represented by the markets) as to where responsibilities for the deployment and use of digital technologies should lie. Responsibility is pushed away from these competing actors and towards the individual technology user who has the least power to push back on the obligations and duties they are ascribed. We suggest that much could be done to better map where digital responsibilities are sited and for what realm of environmental activity. Such mapping could be an extension to existing security practices already included in technology design processes, user walkthroughs and technology testing.

Resource allocation and revenue generation

Our final consideration is the one that is most contentious, but one that we believe is visible from the case studies. It regards the role of resource allocation and revenue generation and the impact they have on responsibility formation. Needless to say, this is deeply political territory and one that we take care in walking through. That said, we believe it is necessary to walk this path as it highlights the need to consider digital responsibility formation as a collaborative and consensual process that involves realising, establishing and actioning responsibility. It is this broader framework that better explains digital responsibilities and assists policy makers involved in digital policy implementation to better understand the complex contours of digital and physical environments, including the actors that govern those spaces and reside in them. Let's start first with governmental resource allocation.

Digital Migration is the clearest example of the impact of limited resource allocation across the three case studies. Even though the importance of digital support for refugees is clearly realised at the supra-jurisdictional level, it remained to be established and actioned through the state at national and local

Figure 6.5: Resources and business models

levels. Formalised resource allocation strategies were lacking and were exacerbated further by the COVID-19 pandemic. The lockdowns strained every facet of UK society, particularly the most vulnerable, who were suddenly required to negotiate complex bureaucracies through purely online services. Much like Digital Census, the responsibility for providing online access and the skills necessary to complete online forms were absorbed by on-the-ground civil society organisations. Resource stretched charities were essentially forced to further take on the practical role of the 'surrogate state' due to the lack of actioning by local authorities and the lack of resource allocation by central government.

Digital Census does not describe the same degree of resource abandonment. There is no doubt that the inclusion of extended language possibilities in the online census was a positive active of resource allocation. It clearly generated excitement among the Kurdish community about the possibilities of being recognised as a part of UK society. We pointed out the limits of educational support which were willingly absorbed by community groups. The limits of resource allocation, in this sense, are more contested especially in the context of the Digital Migration case study. Base resources were allocated, unlike in Digital Migration, and it was the willing ground-up community response that provided the impetus for additional educational services.

Nevertheless, both case studies show the clear relationship between governmental resource allocation and responsibility realisation, establishment and actions. Even if a responsibility is realised and established it does not guarantee that sufficient resources will be allocated for responsibility actions. Going back to our framework, we can say that the discharge of baseline responsibilities through resource allocation required other parties to absorb responsibilities to provide more effective educational and support resources.

As we have noted elsewhere in the book, there is a strong connection between the implementation of digital technology with its distance from physical space and the need for physical, human connection to take responsibility for its use within that space. We do not see these absorption actions as the outcome of responsibilisation strategies, especially in relation to the Digital Census study. Rather, they are response-able outcomes that are guided by a deep understanding of the physical and social environments in which the technology is used, and indeed, from the deep sense of an individual's ability to positively respond to a given situation. Again, we think this re-emphasises the need to better identify and understand lacunas because they are environments in which positive responsibility acts can be realised and actioned. Lacunas may then point to places of positive and beneficial resource allocation for all parties, including government.

Similar considerations appear to arise in the context of revenue generation by private sector actors. Our book has significantly covered the distributed nature of digital responsibility, not least through governance structures involving public and private sector actors. The clearest case study example is, of course, the online safety debates. The fractious nature of the debates can be read in several ways, most notably as the online service provider response to core business model construction and the revenue threats that it could impose.

Chapter 3 highlighted the responsibilisation issues that arose with the attempts by online service providers to make users more responsible for actions related to content moderation. Content moderation thus became a debate about a potential source of consumer harm pitted against the design, management and implementation of digitally safe spaces. The development of the latter, as Chapter 3 noted, would have a direct impact on the root cause of harms engendered as a core part of business models based on supposed free-of-charge services that enable on target advertising related to finely monitored user behaviour. It is at this point that online service provider attempts to avoid responsibility and re-allocate content moderation actions to users become more visible. Key is online service provider attempts to shift the realising of responsibility to discussions about content moderation as a user generated harm rather than a profit motive, business model consideration. It is this attempt to re-shape the pathway of law reform establishment that starts in earnest with the attempted responsibilisation of users as the appropriate point of content moderation. In doing so, the debate started to shift away from the broader structural implication of what online safety reforms were trying to achieve.

Our framework then allows a better understanding of the role of resource allocation and revenue generation in relation to responsibility formation. It is unsurprising that traditional frameworks for understanding legal responsibility, and to a lesser extent, even contemporary frameworks, largely fail to connect resource allocation with responsibility formation. The reason for this, which we note elsewhere in the book, is that ascribed responsibility through law is seen as a closed and unitary process. It is hopefully apparent by now that the interdisciplinary nature of our framework offers a unique perspective to digital responsibility that can consider the role of law in a broader setting than previously envisaged.

The same, we think, can be said for understanding the relationship between revenue generation by private sector parties and responsibility. The case studies show that marketisation of public services and obligations do not entail the eradication or reduction of responsibilities on policy makers or private sector service deliverers. Lacunas cannot simply be avoided by attempting to make users responsible for digital policy outcomes. Avoiding or refusing responsibility actions does not make responsibilities go away. Instead, it shifts them to other parties that may or may not be better placed or resourced to absorb or reallocate responsibilities. These lacunas will keep appearing as they are an unavoidable consequence of the pluralised nature of digital policy implementation. As such, we need a framework that allows conversation to openly develop, including the surfacing of profit motive discussions which are often tacitly baked into business model construction.

Our focus on sites and realms of responsibility realisation and establishment are important to understanding the formation of digital responsibilities. These are the spaces, whether formed or in lacuna state, in which negotiations take place. Realms determine the active stakeholders involved in consensus building and sites surface the many interests and values foregrounded by parties. Our key focus on expanding responsibility analysis to this breadth is to highlight the possibility of negotiated consensus. Non-resolution loops – for example, the situation where a lacuna is not resolved and thus responsibility actions remain unallocated – are expensive for all parties. Understanding responsibility in a broader sense could lead to the reduction of non-resolution loops and provide better implementation strategies for all parties by having clearer and more targeted discussions about realising, establishing and actioning responsibilities. Doing so reduces the need for technical, and indeed, legal, forcing functions as a means of ending non-resolution loops which become harder to impose on users as time moves on.

Conclusion

Chapter 6 outlines the lessons we think can be learned when our digital responsibility framework is considered against the book's case studies. The framework is by necessity broad and we hope this analysis helps to explain why. It provides a necessarily encompassing picture of how responsibilities are realised, established and actioned that better explains the broader implications of digital policy implementation. Other responsibility frameworks tend to focus on actions alone, whereas we believe actions must be understood as attempts to realise responsibility and how those realisations are established. Our different responsibility processes also highlight the distributed nature of responsibility formation across technical, legal and social domains. Lacunas are an important way to understand responsibility distributions because they highlight points of absence in collaboration and consensus. These points of absence are often considered as gaps that need to be filled by forcing functions technical, legal or otherwise. We show that need not be the case.

All these considerations highlight the intrinsic link between resource allocation, revenue generation and responsibility formation, which again is central to understanding the breadth of our framework. Having outlined the lessons learned, we now conclude the book with five understandings to take away.

SEVEN

Five Understandings

Introduction

We conclude this book with a concise summary of some of the understandings that our work generates and a discussion about how these understandings might lead to actions. Our work has emphasised that responsibilities in a digitally mediated environment are reciprocal and relational. It is also a form of responsibility that is multi-dimensional and pluralistic. Each case study shows some of the ways in which contemporary public policy often overlooks this form of responsibility, diminishing the effectiveness of the policy and making vulnerable those the policy seeks to benefit. Each case study also shows how technology is used to enforce responsible actions without recognising the multifaceted nature of digital responsibilities. Each case study illustrates how community actions do the relational and reflexive work necessary to realise and establish responsibilities and in so doing, reframe the use of technology to one of enabling responsibility actions. It is through these case studies that we see many of the ways that positive, enabling approaches to digital responsibilities are fundamental to a digitally democratic society.

These contentions are represented in a high-level view of our framework (see Figure 7.1) that highlights the bases of enabling collaboration and constructing consensus that are vital to understanding its relational components and aspects.

Figure 7.1: The relational aspects of our digital responsibilities framework

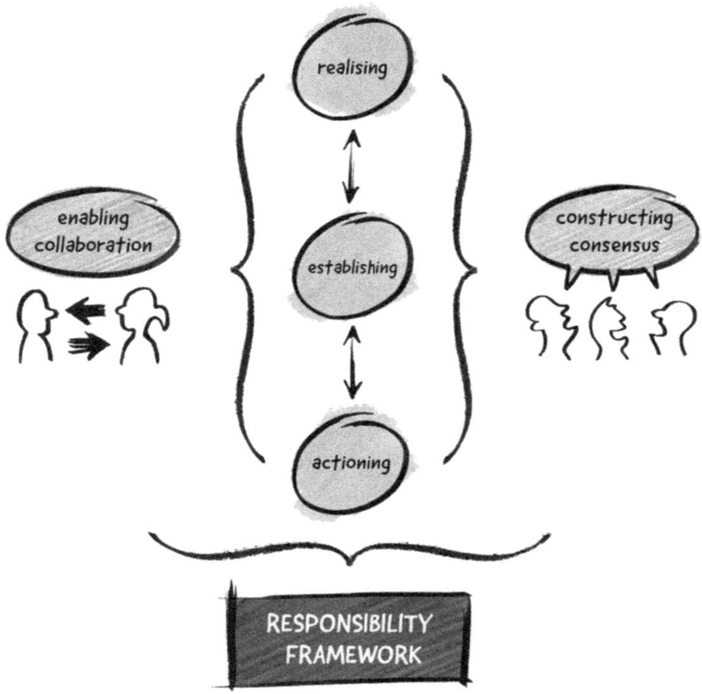

Five understandings

Chapter 1 noted that the RISCS digital responsibility fellowship began with a series of consultations with academic and practitioner communities about what the term 'digital responsibility' meant to them and how the notion of responsibility is shaped by digital technology. The consultations yielded varied, and sometimes conflicted, responses. Nevertheless, a clearer picture of digital responsibilities emerged which eventually crystalised to form our framework. We now put forward five understandings to form the basis of a manifesto that seeks to make conversations about digital responsibilities central to the co-construction of digital policy interactions across time and space.

Understanding the digital is about real relationships

Responsibility can only ever be partially encoded in the design of digital technology and the formulation of law and regulation that frame its design and operation. Our case studies demonstrate that digital responsibilities are also formed through relationships and are negotiated consistently. A focus on relationships is necessary to realise, establish and action responsible use of digital technologies. However, frameworks that seek to encode responsibilities into digital artefacts by themselves, without considering the relationships involved in design and use, will not ensure responsible uses. It is acknowledging the relational dimension of digital responsibilities that enables implementation actions to be adjusted in line with the capabilities and resources of parties. Understanding digital policy implementation thus means identifying and accepting the complexity of digital and physical environments and the relationships that co-exist within them.

Understanding digital impact at scale

Digital technologies re-shape responsibilities in part through their ability to create impact at scale. At the click of a mouse, digital technologies can create impacts that are felt by many over both time and space. This can have the effect of disconnecting the realisation of digital responsibilities from their establishment and from their actioning. Potential dislocation over time and space is another reason why ascribing responsibility into digital technologies is insufficient because implementation impacts are constantly evolving. As the implementation of digital technologies are further honed for scalability and efficiency, potential impacts are further distributed and the allocation of responsibilities become more amorphous. The scale of potential digital impact means that digital responsibilities must be considered holistically, broadly and relationally to ensure that all relevant conversations are considered and enacted.

Understanding that we can't have speed and responsibility

The implementation of digital technologies enables impact at speed, as well as scale. Responsibilities that cannot be encoded into digital technologies impede speed. This raises the especially important question of whether you can have both speed and digital responsibilities formation at the same time. Our case studies indicate that at some point, time and care must be taken by all actors involved in digital policy implementation to converse about which responsibilities should lie where. Our case studies also show that such conversations were largely regarded as a sink cost to be avoided by both policy makers and digital technology manufacturers. Our digital responsibility framework makes visible the conversational work required to generate consensus and acknowledges the efforts of different actors to take the time to have essential responsibility conversations. The conversations are essential to ensure that digital policy implementation is safe and more resilient, which require care, time and patience. Our work then suggests that speed and the processes of responsibility formation are not natural partners.

Understanding responsibility frameworks differently

In contemporary society, diverse groups will understand digital responsibilities in different ways. Hence our framework is a way of thinking about collective actions across different groups and is an intervention to change the conversation about digital responsibilities. The framework offers a means of scaffolding conversations about responsibility in a way that gives structure to a complex, and often contested, topic. From these conversations, groups can start to advocate for certain responsibilities to be actioned and create guidance on how to do it. Our framework is therefore as a means of framing conversations that brings together diverse understandings of digital responsibility rather than being based on the needs and perspectives of one group.

Our framework is developed from the viewpoint that responsibility lacunas will always exist. The temporal and scalar dimensions to digital responsibility are such that there will always be dislocations between realising, establishing and actioning digital responsibilities and these dislocations will give rise to absences and silences. At the same time, the power imbalances between the different stakeholders will also lead to dissensus on where responsibilities lie. Consequently, conversations about digital responsibilities will need to take place at the local as well as the distributed levels, and the framework is designed to support conversations at both scales. All of this means that digital responsibilities need to be considered differently to the traditional approaches of blameworthy ascriptions. Digital responsibilities require consensus, negotiation and collaboration processes which require us to rethink how responsibility has traditionally been conceptualised and ascribed.

Understanding that digital responsibility is pluralistic

Just as digital technologies support distinct types of practice, digital responsibilities can be enacted in many ways. This plurality requires that there be consensus about how digital responsibilities are enacted and about which responsibilities are realised and established. The realisation of digital responsibilities might emerge in one site but equally it could emerge in another. Similarly, establishment could be driven from one realm or equally another. Different actions take place using multiple actors in many different environments. The notion of digital responsibilities is thus deeply pluralistic because it's inherently relational, distributed and fractal, because each of the three processes interact with each other across sites, realms and environments. Hence the need for the plural conversations outlined throughout the book.

Conclusion – understanding that it's different now

In conclusion, we contend that these five understandings shed light on why digital responsibilities are different to the traditional conceptualisations of responsibility. They also make the case for why a relational response to digital responsibilities is needed. These understandings motivate our digital responsibilities framework and shape its intended use.

At the heart of our understandings is the need for multiple, plural and collaborative conversations that are spread over time and space to achieve consensus about digital technology implementation and use. Our understandings have developed over years of thinking and coalesced through writing in 2025. When we started our journey into digital responsibilities, we did not anticipate an unfolding world in which the recognition of plurality, the beneficial acknowledgement of collaboration and the acceptance of consensus as a core democratic function would all be under threat. Yet here we find ourselves.

We appear to be beginning a time where the state use of executive authority is starting to challenge the established democratic order of the second half of the 20th Century. We are also at a time where the power of the largest technology companies is concentrating and becoming increasingly implicated in the use of executive authority. The book's case studies highlight the limits of forcing functions, whether they be in law, or built as security controls in technology. These limits only emanate from the diffuse values and needs of the different sites outlined in Chapter 2, namely, the state, the market and the individual. Responsibilities, including digital responsibilities, will always be contentious as long as the values and needs of the state and the market remain separate, even in societies governed by privatisation or marketisation policies. Should those values and needs be conjoined, then the ability to forcibly ascribe responsibilities on individuals through digital technologies, predicated on responsibilisation strategies, will increase manifestly.

We end our book then with this message.

The world is indeed different now to when we started this work. We started understanding digital responsibilities to better assist with the implementation of digital policy strategies so all parties could collaborate conversationally for the betterment of all. We still believe that it is essential to understand the varied and complex responsibilities that arise and which are distributed over time, location and space due to digital technology. The distributed nature of digital responsibility formation is itself a protection. Our digitally mediated worlds are expansive, messy and everywhere. Understanding digital responsibilities therefore becomes vital to better see the underlying benefits and dangers of digital-by-default societies, especially at a time when the foundations of governmental and technology architectures are changing rapidly. Going back to Peter Cane in Chapter 2, understanding who we are responsible to, as well as what we are responsible for, has never been more important than it is now, especially in the digital context. Truly understanding digital responsibilities, therefore, is as much grounded in the physical world as it is in the digital one.

References

AbuJarour, S., Krasnova, H., Andrade, A., Olbrich, S., Tan, C.W., Urquhart, C., Wiesche, M., 2017. Empowering Refugees with Technology: Best Practices and Research Agenda, in: Proceedings of the 25th European Conference on Information Systems. Presented at the ECIS 2017.

Acker, A., 2025. Archiving Machines. MIT Press.

Advertising Standards Authority, 2020. Protecting Young and Vulnerable People.

Ahmed, B., 2018. Who takes responsibility for the climate refugees? International Journal of Climate Change Strategies and Management 10, 5–26. https://doi.org/10.1108/IJCCSM-10-2016-0149

Ajana, B., 2013. Asylum, identity management and biometric control. Journal of Refugee Studies 26, 576–95. https://doi.org/10.1093/jrs/fet030

Alencar, A., Camargo, J., 2023. Spatial imaginaries of digital refugee livelihoods. Journal of Humanitarian Affairs, 22–30.

Ali, A., Champion, N., 2021. More harm than good: A super-complaint on the harms caused by 'suspicion-less' stop and searches and inadequate scrutiny of stop and search powers. Criminal Justice Alliance.

Almohamed, A., Vyas, D., 2016. Vulnerability of Displacement: Challenges for Integrating Refugees and Asylum Seekers in Host Communities, in: Proceedings of the 28th Australian Conference on Computer-Human Interaction, 125–34. https://doi.org/10.1145/3010915.3010948

Angeloni, C., 2022. UK to tackle 'epidemic' of paid-for scam adverts. International Adviser. URL https://international-adviser.com/uk-to-tackle-epidemic-of-paid-for-scam-adverts/ (accessed 3.20.25).

Barnardo's, 2021. Barnardo's survey reveals the harm done to children by harmful pornography – Culture Sec must. Barnardo's. URL https://www.barnardos.org.uk/news/barnardos-survey-reveals-harm-done-children-harmful-pornography-culture-sec-must (accessed 3.20.25).

Barnes, S.B., 2000. Understanding Agency: Social Theory and Responsible Action. SAGE Publications, Limited.

Barton, C., House of Commons Library, 2021. Preparing for the 2021 Census (England and Wales). House of Commons Library.

Baser, B., 2011. Kurdish diaspora political activism in Europe with a particular focus on Great Britain. Berghof Peace Support Centre for Just Peace and Democracy.

BBC, 2021. Online Safety Bill: new offences and tighter rules. BBC. URL https://www.bbc.co.uk/news/technology-59638569 (accessed 9.4.25).

Bearitz, K., Schertel Mendes, L., 2023. A primer on the UK Online Safety Act. Verfassungsblog. https://doi.org/10.59704/2120f79b5f59e60b

Bellini, R., Forrest, S., Westmarland, N., Smeddinck, J.D., 2020. Mechanisms of Moral Responsibility: Rethinking Technologies for Domestic Violence Prevention Work, in: Proceedings of the 2020 CHI Conference on Human Factors in Computing Systems. https://doi.org/10.1145/3313831.3376693

Boehner, K., DiSalvo, C., 2016. Data, Design and Civics: An Exploratory Study of Civic Tech, in: 2016 CHI Conference on Human Factors in Computing Systems. Presented at the CHI '16, 2970–81. https://doi.org/10.1145/2858036.2858326

Bogard, W., 2005. Welcome to the Society of Control: The Simulation of Surveillance Revisited. University of Toronto Press.

Bozalek, V., Zemblylas, M., 2023. Responsibility, Privileged Irresponsibility and Response-ability: Higher Education, Coloniality and Ecological Damage. Palgrave Critical University Studies. Palgrave Macmillan.

Burns, H., 2021a. The UK's Online Safety Bill is a 'hostage-taking' law. That should Terrify you. Hi, I'm Heather Burns. URL https://heatherburns.tech/2021/11/08/online-safety-bill-hostage-taking-law/ (accessed 3.20.25).

Burns, H., 2021b. Age verification in the Online Safety Bill. Open Rights Group. URL https://www.openrightsgroup.org/blog/age-verification-in-the-online-safety-bill/ (accessed 3.20.25).

Cane, P., 2016. Role responsibility. The Journal of Ethics 20, 279–98. https://doi.org/10.1007/s10892-016-9235-8

Cane, P., 2002. Responsibility in Law and Morality. Hart Publishing. https://doi.org/10.5040/9781472562531

Cane, P., 2001. Responsibility and fault: a relational and functional approach to responsibility, in: Relating to Responsibility: Essays in Honour of Tony Honoré on His 80th Birthday. Hart Publishing, pp 81–110.

Castells, M., 2009. Communication Power. Oxford University Press.

Chouhan, C., LaPerriere, C.M., Aljallad, Z., Kropczynski, J., Lipford, H., Wisniewski, P.J., 2019. Co-designing for Community Oversight. Proceedings of the ACM on Human-Computer Interaction 3, 1–31. https://doi.org/10.1145/3359248

Cipolla, C., Manzini, E., 2009. Relational services. Knowledge, Technology & Policy 22, 45–50. https://doi.org/10.1007/s12130-009-9066-z

Coe, P., 2022. The Draft Online Safety Bill and the regulation of hate speech: have we opened Pandora's box? Journal of Media Law 14, 50–75. https://doi.org/10.1080/17577632.2022.2083870

Coeckelbergh, M., 2023. Narrative responsibility and artificial intelligence. AI & Society 38, 2437–50. https://doi.org/10.1007/s00146-021-01375-x

Coeckelbergh, M., 2020. Artificial intelligence, responsibility attribution, and a relational justification of explainability. Science and Engineering Ethics 26, 2051–68. https://doi.org/10.1007/s11948-019-00146-8

Coles-Kemp, L., Hall, P.A., 2023. Digital security narratives in the time of COVID-19: a case for kindness, in: Martins, N., Brandão, D., Paiva, F. (Eds.), Perspectives on Design and Digital Communication III: Research, Innovations and Best Practices. Springer International Publishing, pp 123–35. https://doi.org/10.1007/978-3-031-06809-6_8

Coles-Kemp, L., Jensen, R.B., 2019. Accessing a New Land: Designing for a Social Conceptualisation of Access, in: Proceedings of the 2019 CHI Conference on Human Factors in Computing Systems (CHI '19). Association for Computing Machinery, New York, NY, USA. Paper 181, 1–12. https://doi.org/10.1145/3290605.3300411

Coles-Kemp, L., Bjerg Jensen, R., Heath, C.P., 2020. Too Much Information: Questioning Security in a Post-Digital Society, in: Proceedings of the 2020 CHI Conference on Human Factors in Computing Systems. Presented at the CHI '20, Hawaii. https://doi.org/10.1145/3313831.3376214

Coles-Kemp, L., Robinson, N., Heath, C.P., 2022. Protecting the Vulnerable: Dimensions of Assisted Digital Access. Proceedings of the ACM on Human-Computer Interaction 6, 1–26.

Commission: Directorate-General for Research and Innovation, Schomberg, R. v., 2011. Towards responsible research and innovation in the information and communication technologies and security technologies fields. Publications Office. https://doi.org/10.2777/58723

Cookson, R., 2016. 'Endemic' fraud threatens digital advertising budgets. Financial Times. URL https://www.ft.com/content/8e2f59c0-2b01-11e6-a18d-a96ab29e3c95 (accessed 3.20.25).

Dancohen, M., 1992. Responsibility and the boundaries of the self. Harvard law Review 105, 959–1003. https://doi.org/10.2307/1341517

Davies, W., 2018. The neoliberal state: power against 'politics', in: Cahill, D., Cooper, M., Konings, M., Primrose, D. (Eds.), The SAGE Handbook of Neoliberalism. SAGE Publications, pp 273–83.

Deardorff Miller, S., 2017. UNHCR as a Surrogate State: Protracted Refugee Situations. Routledge.

Deleuze, G., 1992. Postscript on the societies of control. October 59, 3–7.

Demir, I., 2012. Battling with Memleket in London: the Kurdish diaspora's engagement with Turkey. Journal of Ethnic and Migration Studies 38, 815–31. https://doi.org/10.1080/1369183x.2012.667996

Department for Digital, Culture, Media & Sport, 2022. World-leading measures to protect children from accessing pornography online. GOV.UK. URL https://www.gov.uk/government/news/world-leading-measures-to-protect-children-from-accessing-pornography-online (accessed 3.20.25).

Department for Digital, Culture, Media & Sport, 2017. Internet Safety Strategy green paper. GOV.UK. URL https://www.gov.uk/government/consultations/internet-safety-strategy-green-paper (accessed 3.20.25).

Department for Digital, Culture, Media & Sport, Home Office, 2019. Online Harms White Paper. GOV.UK. URL https://www.gov.uk/government/consultations/online-harms-white-paper/online-harms-white-paper (accessed 3.20.25).

Dignum, V., 2020. Responsibility and artificial intelligence, in: The Oxford Handbook of Ethics of AI. Oxford University Press. https://doi.org/10.1093/oxfordhb/9780190067397.013.12 (accessed 9.11.25).

Dignum, V., 2019. Responsible Artificial Intelligence: How to Develop and Use AI in a Responsible Way. Artificial Intelligence: Foundations, Theory, and Algorithms. Springer. https://doi.org/10.1007/978-3-030-30371-6

Dorr, S., 2021. Corporate Digital Responsibility. Springer.

Dowd, R., McAdam, J., 2017. International cooperation and responsibility-sharing to protect refugees: What, why and how? International and Comparative Law Quarterly 66(4), 863–92.

Dunn Cavelty, M., Egloff, F.J., 2019. The politics of cybersecurity: balancing different roles of the state. St Antony's International Review 15, 37–57.

Easton-Calabria, E., Allen, W.L., 2015. Developing ethical approaches to data and civil society: from availability to accessibility. Innovation: The European Journal of Social Science Research 28, 52–62. https://doi.org/10.1080/13511610.2014.985193

Elliott, K., Price, R., Shaw, P., Spiliotopoulos Tasos, Ng, M., Kovila, C., Aad, van M., 2021. Towards an equitable digital society: artificial intelligence (AI) and corporate digital responsibility (CDR). Society 58, 179–88. https://doi.org/10.1007/s12115-021-00594-8

Enneli, P., Modood, T., Bradley, H., 2005. Young Turks and Kurds: A Set of 'Invisible' Disadvantaged Groups. University of Bristol Joseph Rowntree Foundation.

Fan, B., Fox, S.E., 2022. Access under duress: pandemic-era lessons on digital participation and datafication in civic engagement. Proceedings of the ACM on Human-Computer Interaction 6, 1–22. https://doi.org/10.1145/3492833

Faulkner, D., Lee, J., 2022. UK Rwanda asylum plan against international law, says UN refugee agency. BBC News. URL https://www.bbc.com/news/uk-61122241 (accessed 3.20.25).

Fischer, J.M., Ravizza, M., 1998. Responsibility and Control: A Theory of Moral Responsibility, Cambridge Studies in Philosophy and Law. Cambridge University Press. https://doi.org/10.1017/CBO9780511814594

Fisher, K.E., Yafi, E., 2018. Syrian Youth in Za'atari Refugee Camp as ICT Wayfarers: An Exploratory Study Using LEGO and Storytelling, in: Proceedings of the 1st ACM SIGCAS Conference on Computing and Sustainable Societies, 32, 1–12. https://doi.org/10.1145/3209811.3209873

Fisher, K.E., Yefimova, K., Yafi, E., 2016. Future's Butterflies: Co-Designing ICT Wayfaring Technology with Refugee Syrian Youth, in: Proceedings of the 15th International Conference on Interaction Design and Children (IDC '16). Association for Computing Machinery, New York, NY, USA, 25–36. https://doi.org/10.1145/2930674.2930701

Frouws, W.B., Phillips, M., Hassan, A., Twigt, M., 2016. Getting to Europe the 'WhatsApp' Way. Danish Refugee Council.

Gardner, J., 2003. The mark of responsibility. Oxford Journal of Legal Studies 23, 157–71. https://doi.org/10.1093/ojls/23.2.157

Giddens, A., 1991. Modernity and Self-identity: Self and Society in the Late Modern Age. Stanford University Press.

Glick, P., Crivellaro, C., 2023. MyCareBudget: co-creating a healthcare digital commons with and for disabled citizens and their unpaid carers. ACM, pp 1–16. https://doi.org/10.1145/3544548.3580934

Gopal, D., Matras, Y., 2013. What languages are spoken in England and Wales? Dynamic of Diversity: Evidence from the 2011 Census. ESRC Centre on Dynamics of Ethnicity (CoDE), The University of Manchester. URL https://hummedia.manchester.ac.uk/institutes/code/briefingsupdated/what-languages-are-spoken-in-england-and-wales.pdf (accessed 9.17.25).

Gough, H.A., Gough, K.V., 2019. Disrupted becomings: the role of smartphones in Syrian refugees' physical and existential journeys. Geoforum 105, 89–98. https://doi.org/10.1016/j.geoforum.2019.05.012

Gunkel, D.J., 2020. Mind the gap: responsible robotics and the problem of responsibility. Ethics and Information Technology 22, 307–20. https://doi.org/10.1007/s10676-017-9428-2

Hart, H.L.A., 2000. Legal responsibility and excuses, in: Coleman, J.L. (Ed.), Readings in the Philosophy of Law. Taylor & Francis, pp 495–520.

Hart, H.L.A., 1968. Punishment and Responsibility: Essays in the Philosophy of Law, [Rev. ed.]. Clarendon Press.

Hart, H.L.A., 1949. XI. – The ascription of responsibility and rights. Proceedings of the Aristotelian Society 49, 171–94. https://doi.org/10.1093/aristotelian/49.1.171

Honore, T., 1999. Responsibility and Fault. Hart Publishing. https://doi.org/10.5040/9781472561954

Hurwitz, A., 2009. The Collective Responsibility of States to Protect Refugees. Oxford University Press. https://doi.org/10.1093/acprof:oso/9780199278381.001.0001

Information Commissioner's Office, 2020. Age-appropriate design: a code of practice for online services. https://ico.org.uk/for-organisations/uk-gdpr-guidance-and-resources/childrens-information/childrens-code-guidance-and-resources/age-appropriate-design-a-code-of-practice-for-online-services/ (accessed 3.20.25).

International Committee of the Red Cross, 2024. Handbook on data protection in humanitarian action. International Committee of the Red Cross. URL https://www.icrc.org/en/data-protection-humanitarian-action-handbook (accessed 3.20.25).

Internet Society, 2021. Internet Society: UK Online Public Safety Bill is trying to legislate the impossible – a safe Internet without strong encryption. Internet Society. URL https://www.internetsociety.org/news/statements/2021/internet-society-uk-online-public-safety-bill-is-trying-to-legislate-the-impossible-a-safe-internet-without-strong-encryption/ (accessed 3.20.25).

Jacobsen, K.L., Sandvik, K.B., 2018. UNHCR and the pursuit of international protection: accountability through technology? Third World Quarterly 39, 1508–24. https://doi.org/10.1080/01436597.2018.1432346

Jensen, R.B., Coles-Kemp, L., Talhouk, R., 2020a. When the Civic Turn turns Digital: Designing Safe and Secure Refugee Resettlement, in: Proceedings of the 2020 CHI Conference on Human Factors in Computing Systems. Presented at the CHI '20. https://doi.org/10.1145/3313831.3376245

Jensen, R.B., Coles-Kemp, L., Wendt, N., Lewis, M., 2020b. Digital Liminalities: Understanding Isolated Communities on the Edge, in: ACM CHI Conference on Human Factors in Computing Systems. DOI:10.1145/3313831.3376137

Jones, A.E., 2013. Exploring teacher perceptions and expectations of Turkish, Kurdish and Turkish Cypriot girls from the perspectives of both teachers and students in three London secondary schools. Race Ethnicity and Education 17, 667–87. https://doi.org/10.1080/13613324.2012.759929

Jongerden, J., Akkaya, A.H., 2018. The Kurdistan Workers Party (PKK) and Kurdish Political Parties in the 1970s, in: Routledge Handbook on the Kurds. Routledge.

Krasnova, H., AbuJarour, S., Fedorowicz, J., Olbrich, S., Tan, C.W., Urquhart, C., Venkatesh, V., 2016. Leveraging Technology for Refugee Integration: How Can We Help?, in: Proceedings of the International Conference on Information Systems. Presented at the ICIS 2016 Proceedings, Dublin, Ireland.

Kutz, C., 2004. 548 Responsibility, in: Coleman, J.L., Himma, K.E., Shapiro, S.J. (Eds.), The Oxford Handbook of Jurisprudence and Philosophy of Law. Oxford University Press, pp 548–87. https://doi.org/10.1093/oxfordhb/9780199270972.013.0014

Latonero, M., Kift, P., 2018. On digital passages and borders: refugees and the new infrastructure for movement and control. Social Media + Society 4, 2056305118764432. https://doi.org/10.1177/2056305118764432

Latonero, M., Hiatt, K., Napolitano, A., Clericetti, G., Penagos, M., 2019. Digital Identity in the Migration & Refugee Context: Italy Case Study. Data & Society.

Le Dantec, C., 2016a. Designing Publics. MIT Press.

Le Dantec, C., 2016b. Design through collective action / collective action through design. Interactions 24, 24–30. https://doi.org/10.1145/3018005

Leiser, M.R., Harbinja, E., 2020. Content Not Available: Why The United Kingdom's Proposal for A 'Package of Platform Safety Measures' Will Harm Free Speech. TechReg 2020, 78–90. https://doi.org/10.26116/techreg.2020.008

Lima, G., Cha, M., 2020. Responsible AI and its Stakeholders. https://doi.org/10.48550/arXiv.2004.11434

Liyanaarachchi, G., Deshpande, S., Weaven, S. 2021. Market-oriented corporate digital responsibility to manage data vulnerability in online banking. International Journal of Bank Marketing 39, 4, 571–91. https://doi.org/10.1108/IJBM-06-2020-0313

Lobschat, L., Mueller, B., Eggers, F., Brandimarte, L., Diefenbach, S., Kroschke, M., Wirtz, J., 2021. Corporate digital responsibility. Journal of Business Research 122, 875–88. https://doi.org/10.1016/j.jbusres.2019.10.006

Lynn, T., Rosati, P., Conway, E., Curran, D., Fox, G., O'Gorman, C., 2022. Digital Technologies and Civil Society, in: Digital Towns: Accelerating and Measuring the Digital Transformation of Rural Societies and Economies. Springer International Publishing, pp 91–108. https://doi.org/10.1007/978-3-030-91247-5_5

Markus, M.L., Mentzer, K., 2014. Foresight for a responsible future with ICT. Information Systems Frontiers 16, 353–68. https://doi.org/10.1007/s10796-013-9479-9

Martin, S.F., Davis, R., Benton, G., Waliany, Z., 2018. International Responsibility-Sharing for Refugees, KNOMAD Working Paper. World Bank.

Matthias, A., 2004. The responsibility gap: ascribing responsibility for the actions of learning automata. Ethics and Information Technology 6, 175–83. https://doi.org/10.1007/s10676-004-3422-1

McDowall, D., 2021. A modern history of the Kurds, 4th edn. I.B. Tauris.

McGlynn, C., Woods, L., Antoniou, A., 2024. Pornography, the Online Safety Act 2023 and the need for further reform. Journal of Media Law 1–29. https://doi.org/10.1080/17577632.2024.2357421

Nash, V., 2019. Revise and resubmit? Reviewing the 2019 Online Harms White Paper. Journal of Media Law 11, 18–27. https://doi.org/10.1080/17577632.2019.1666475

Nash, V., Felton, L., 2024. Treating the symptoms or the disease? Analysing the UK Online Safety Act's approach to digital regulation. Policy & Internet 16, 4, 818–32. https://doi.org/10.1002/poi3.404

Ognjenovic, G., 2010. Responsibility in Context Perspectives. Springer Netherlands. https://doi.org/10.1007/978-90-481-3037-5

ONS, 2020. Ethnic group tick-box prioritisation report for Census 2021 in England and Wales. ONS. URL https://www.ons.gov.uk/census/censustransformationprogramme/questiondevelopment/ethnicgrouptickboxprioritisationreportforcensus2021inengland andwales (accessed 9.5.25).

ONS, 2011. 2011 Census for England and Wales. ONS. URL https://www.ons.gov.uk/census/2011census (accessed 9.5.25).

Peeters, R., 2013. Responsibilisation on Government's terms: new welfare and the governance of responsibility and solidarity. Social Policy and Society 12, 583–95. https://doi.org/10.1017/S14747 46413000018

Perrin, W., Woods, L., 2018. Harm Reduction in Social Media – A Proposal. URL https://carnegieuk.org/blog/harm-reduction-in-social-media-a-proposal/ (accessed 3.20.25).

Porter, M., Haslam, N., 2005. Predisplacement and postdisplacement factors associated with mental health of refugees and internally displaced persons: a meta-analysis. JAMA: The Journal of the American Medical Association 294, 602–12. https://doi.org/10.1001/jama.294.5.602

Price, L., 2021. Platform responsibility for online harms: towards a duty of care for online hazards. Journal of Media Law 13, 238–61. https://doi.org/10.1080/17577632.2021.2022331

Renaud, K., Flowerday, S., Warkentin, M., Cockshott, P., Orgeron, C., 2018. Is the responsibilization of the cyber security risk reasonable and judicious? Computers & Security 78, 198–211. https://doi.org/10.1016/j.cose.2018.06.006

Roe, P., 2008. The 'value' of positive security. Review of International Studies 34, 777–94. doi:10.1017/S0260210508008279

Rose, N., 2000. Government and control. The British Journal of Criminology 40, 321–39. https://doi.org/10.1093/bjc/40.2.321

Ruiu, M.L., Ragnedda, M., 2016. Between digital inclusion and social equality: the role of public libraries in Newcastle upon Tyne. Library and Information Research 40, 69–87. https://doi.org/10.29173/lirg715

Scanlon, T., 2000. What We Owe to Each Other. Belknap Press of Harvard University Press.

Schoemaker, E., Currion, P., Pon, B., 2018. Identification Systems for Refugees. Caribou Digital.

Selwyn, N., Gorard, S., Furlong, J., 2005. Whose internet is it anyway? Exploring adults' (non)use of the internet in everyday life. European Journal of Communication 20, 5–26.

Sentas, V., 2018. Terrorist organization proscription as counter-insurgency in the Kurdish conflict. Terrorism and Political Violence 30, 298–317. https://doi.org/10.1080/09546553.2018.1432215

Sentas, V., 2016. Policing the diaspora: Kurdish Londoners, MI5 and the proscription of terrorist organizations in the United Kingdom. British Journal of Criminology 56, 898–918. https://doi.org/10.1093/bjc/azv094

Shamir, R., 2008. The age of responsibilization: on market-embedded morality. Economy and Society 37, 1–19. https://doi.org/10.1080/03085140701760833

Simko, L., Lerner, A., Ibtasam, S., Roesner, F., Kohno, T., 2018. Computer Security and Privacy for Refugees in the United States, in: 2018 IEEE Symposium on Security and Privacy (SP), pp 409–23. https://doi.org/10.1109/SP.2018.00023

Sinayic, B., 2021. Count Me In: Census 2021. YouTube. URL https://www.youtube.com/watch?v=2OXURUl4_VQ (accessed 9.17.25).

Sirkeci, I., 2000. Exploring the Kurdish population in the Turkish context. Genus 56, 149–75.

Smith, G., 2020. An Online Harms Compendium. Cyberleagle. URL https://www.cyberleagle.com/2020/02/an-online-harms-compendium.html (accessed 3.20.25).

Stilgoe, J., Owen, R., Macnaghten, P., 2013. Developing a framework for responsible innovation. Research Policy 42, 1568–80. https://doi.org/10.1016/j.respol.2013.05.008

Talhouk, R., Mesmar, S., Thieme, A., Balaam, M., Olivier, P., Akik, C., Ghattas, H., 2016. Syrian Refugees and Digital Health in Lebanon: Opportunities for Improving Antenatal Health, in: Proceedings of the 2016 CHI Conference on Human Factors in Computing Systems (CH '16). Association for Computing Machinery, New York, NY, USA, 331–42. https://doi.org/10.1145/2858036.2858331

Tambini, D., 2019. The differentiated duty of care: a response to the Online Harms White Paper. Journal of Media Law 11, 28–40. https://doi.org/10.1080/17577632.2019.1666488

Tigard, D.W., 2021. There is no techno-responsibility gap. Philos. Technol. 34, 589–607. https://doi.org/10.1007/s13347-020-00414-7

Topak, Ö.E., 2014. The biopolitical border in practice: surveillance and death at the Greece-Turkey borderzones. Environ Plan D 32, 815–33. https://doi.org/10.1068/d13031p

Trengove, M., Kazim, E., Almeida, D., Hilliard, A., Zannone, S., Lomas, E., 2022. A critical review of the Online Safety Bill. Patterns 3. https://doi.org/10.1016/j.patter.2022.100544

Trnka, S., Trundle, C., 2014. Competing responsibilities: moving beyond neoliberal responsibilisation. Anthropological Forum 24, 136–53. https://doi.org/10.1080/00664677.2013.879051

Udwan, G., Leurs, K., Alencar, A., 2020. Digital resilience tactics of Syrian refugees in the Netherlands: social media for social support, health, and identity. Social Media + Society 6, 2056305120915587. https://doi.org/10.1177/2056305120915587

UK Parliament, 2022. The Draft Online Safety Bill and the legal but harmful debate – Digital, Culture, Media and Sport Committee. UK Parliament. URL https://publications.parliament.uk/pa/cm5802/cmselect/cmcumeds/1039/report.html (accessed 3.20.25).

Ungar, M., Seymour, A., 2024. Access without borders: a scoping review to identify solutions to creating portable identity, education and health records for refugee children. Journal of International Migration and Integration 25, 1989–2017. https://doi.org/10.1007/s12134-024-01156-7

UNHCR, 2023. Global Trends Report. UNHCR. URL https://www.unhcr.org/global-trends-report-2023 (accessed 3.20.25).

UNHCR, 2022a. Connectivity for Refugees. UNHCR. URL https://www.unhcr.org/innovation/connectivity-for-refugees/ (accessed 3.20.25).

UNHCR, 2022b. UNHCR Strategy on Digital Identity and Inclusion. UNHCR. URL https://www.unhcr.org/blogs/wp-content/uploads/sites/48/2018/03/2018-02-Digital-Identity_02.pdf (accessed 9.17.25).

UNHCR, 2022c. Connected Education for Refugees: Addressing the Digital Divide. UNHCR. URL https://www.unhcr.org/media/connected-education-refugees-addressing-digital-divide (accessed 3.20.25).

UNHCR, 2022d. Communicating with Communities. UNHCR Innovation. URL https://www.unhcr.org/innovation/communicating-with-communities/ (accessed 3.20.25).

UNHCR, 2022e. Asylum in the UK. UNHCR UK. URL https://www.unhcr.org/uk/asylum-uk (accessed 3.20.25).

UNHCR, 2020a. Collaboration for Connectivity. UNHCR. URL https://www.unhcr.org/innovation/wp-content/uploads/2020/06/Collaborating-for-Connectivity_WEB062020.pdf (accessed 9.17.25).

UNHCR, 2020b. Space and Imagination: Rethinking Refugees' Digital Access. UNHCR Innovation Service.

UNHCR, 2018a. Bridging the identity divide – Is Portable User-Centric Identity Management the Answer? UNHCR Blog. URL https://www.unhcr.org/blogs/bridging-identity-divide-portable-user-centric-identity-management-answer/ (accessed 3.20.25).

UNHCR, 2018b. From proGres to PRIMES. UNHCR. URL https://www.unhcr.org/blogs/wp-content/uploads/sites/48/2018/03/2018-03-16-PRIMES-Flyer.pdf (accessed 9.17.25).

UNHCR, 2017. ID2020 and UNHCR Host Joint Workshop on Digital Identity. UNHCR Blog. URL https://www.unhcr.org/blogs/id2020-and-unhcr-host-joint-workshop-on-digital-identity/ (accessed 3.20.25).

United Nations, 2018. Global Compact on Refugees. United Nations. URL https://globalcompactrefugees.org/sites/default/files/2019-12/Global%20compact%20on%20refugees%20EN.pdf (accessed 9.17.25).

van Bruinessen, M., 2000. Turkey and the Kurds in the early 1990s: Guerrilla, counter-insurgency, and emerging civil society, in: Kurdish Ethno-Nationalism versus Nation-Building States. The ISIS Press.

van de Poel, I., 2011. The relation between forward-looking and backward-looking responsibility, in: Vincent, N.A., van de Poel, I., van den Hoven, J. (Eds.), Moral Responsibility: Beyond Free Will and Determinism. Springer Netherlands, pp 37–52. https://doi.org/10.1007/978-94-007-1878-4_3

van de Poel, I.R., Sand, M., 2018. Varieties of responsibility: two problems of responsible innovation. Synthese (Dordrecht) 198, 4769–87. https://doi.org/10.1007/s11229-018-01951-7

van Wilgenburg, W., 2021. Kurdish organizations call on Kurds in UK to register for upcoming census. Kurdistan24.

Vlachokyriakos, V., Crivellaro, C., Le Dantec, Christopher A., Gordon, E., Wright, P., Olivier, P., 2016. Digital civics: citizen empowerment with and through technology, in: Proceedings of the 2016 CHI Conference Extended Abstracts on Human Factors in Computing Systems. Presented at the CHI EA'16, pp 1096–99.

Vrânceanu, A., Dinas, E., Heidland, T., Ruhs, M., 2023. The European refugee crisis and public support for the externalisation of migration management. European Journal of Political Research 62, 1146–67. https://doi.org/10.1111/1475-6765.12565

Wahlbeck, O., 1998. Community work and exile politics: Kurdish refugee associations in London. Journal of Refugee Studies 11, 215–30. https://doi.org/10.1093/jrs/11.3.215

Wahome, M., Graham, M., 2020. Spatially shaped imaginaries of the digital economy. Information, Communication & Society 23, 1123–38. https://doi.org/10.1080/1369118X.2019.1701696

Woods, L., 2019. The duty of care in the Online Harms White Paper. Journal of Media Law 11, 6–17. https://doi.org/10.1080/17577632.2019.1668605

Younus, M., Qodir, Z., Nurmandi, A., Mutiarin, D., Prianto, A.L., Zamjanah Rahmawati, I., Amiruddin, I., Khaerunnisaa, N.I., Harakan, A., Tahir, N., Vhilbar, Z., 2024. Redefining citizenship in the digital age: understanding smart digital census for citizen's population data collection and analysis. International Journal on Recent and Innovation Trends in Computing and Communication 12, 1–9.

Završnik, A., 2021. Algorithmic justice: algorithms and big data in criminal justice settings. European Journal of Criminology 18, 623–42. https://doi.org/10.1177/1477370819876762

Joint Committee Evidence Cited in Chapter 3

JC-1. "Draft Online Safety Bill: Oral evidence – Imran Ahmed, CEO and Founder, Center for Countering Digital Hate."

JC-2. "Draft Online Safety Bill: Oral evidence – Sanjay Bhandari, Chair, Kick It Out; Edleen John; Director of International Relations, Corporate Affairs and Co-Partner for Equality, Diversity and Inclusion, Football Association; Rio Ferdinand."

JC-3. "Draft Online Safety Bill: Oral evidence – Danny Stone MBE, Chief Executive, Antisemitism Policy Trust; Nancy Kelley, Chief Executive, Stonewall."

JC-4. "Draft Online Safety Bill: Oral evidence – Clare Pelham, Chief Executive, Epilepsy Society; Matt Harrison, Public Affairs and Parliamentary Manager, Royal Mencap Society; Nina Jankowicz, Author and Global Fellow, Wilson Center; Izzy Wick, Director of UK Policy, 5Rights Foundation; Ian Russell, Chair of Trustees, Molly Rose Foundation".

JC-5. "Draft Online Safety Bill: Oral evidence – William Perrin OBE, Trustee, Carnegie UK Trust; Dr Edina Harbinja, Senior Lecturer in Media and Privacy Law at Aston University; Professor Sonia Livingstone, LSE Department of Media and Communications; Professor Clare McGlynn, Professor of Law, University of Durham."

JC-6. "Draft Online Safety Bill: Oral evidence – Jimmy Wales, Founder, Wikipedia."

JC-7. "Draft Online Safety Bill: Oral evidence – "Elizabeth Denham CBE, Information Commissioner; Stephen Bonner, Executive Director, Regulatory Futures & Innovation, Information Commissioner's Office."

JC-8. "Draft Online Safety Bill: Oral evidence – Guillaume Chaslot, Founder, Algo Transparency; Rene DiResta, Research Manager, Stanford Internet Observatory; Laura Edelson, Researcher, New York University."

JC-9. "Draft Online Safety Bill: Oral evidence – Martin Lewis, Founder and Chair of MoneySavingExpert.com and the Mental Health and Money Policy Institute; Rocio Concha, Director of Policy and Advocacy and Chief Economist, Which?."

JC-10. "Draft Online Safety Bill: Oral evidence – Mark Steward, Executive Director of Enforcement and Market Oversight, Financial Conduct Authority; Guy Parker, Chief Executive, Advertising Standards Authority; Michael Grenfell, Executive Director for Enforcement, Competition and Markets Authority; T/Commander Clinton Blackburn, National Economic Crime Coordinator, City of London Police".

JC-11. "Draft Online Safety Bill: Oral evidence – Sophie Zhang, former Facebook employee."

JC-12. "Draft Online Safety Bill: Oral evidence – Professor Richard Wilson, Associate Dean for Faculty Development and Intellectual Life, Gladstein Chair and Professor of Anthropology and Law, University of Connecticut (appearing virtually by Zoom); Barbora Bukovská, Senior Director, Law and Policy, Article 19; Matthew d'Ancona, journalist, formerly of Index on Censorship, currently Editor at Tortoise Media (appearing virtually via Zoom); Silkie Carlo, Director, Big Brother Watch."

JC-13. "Draft Online Safety Bill: Oral evidence – Gavin Millar QC, Specialist in Media Law, Matrix; Alison Gow, President, Society of Editors (appearing virtually via Zoom); Matt Rogerson, Director of Public Policy, Guardian Media Group, and News Media Association; Peter Wright, Editor Emeritus, DMG Media".

JC-14. "Draft Online Safety Bill: Oral evidence – Professor Jonathan Haidt; Jim Steyer, Chief Executive Officer, Common Sense Media."

JC-15. "Draft Online Safety Bill: Oral evidence – Frances Haugen, former Facebook employee."

JC-16. "Draft Online Safety Bill: Oral evidence – Maria Ressa, CEO, Rappler."

JC-17. "Draft Online Safety Bill: Oral evidence – Antigone Davis, Global Head of Safety at Facebook; Chris Yiu, Director of Public Policy for Northern Europe at Facebook."

JC-18. "Draft Online Safety Bill: Oral evidence – Leslie Miller, Vice President, Government Affairs and Public Policy, YouTube; Markham C. Erickson, Vice President, Government Affairs and Public Policy, Google."

JC-19. "Draft Online Safety Bill: Oral evidence – Nick Pickles, Senior Director, Global Public Policy Strategy, Development and Partnerships, Twitter; Dr Theo Bertram, Director of Government Relations, Europe, TikTok."

JC-20. "Draft Online Safety Bill: Oral evidence – Dame Melanie Dawes DCB, Chief Executive, Ofcom; Richard Wronka, Director for Online Harms Ofcom."

JC-21. "Draft Online Safety Bill: Oral evidence – Rt Hon Nadine Dorries MP, Secretary of State for Digital, Culture, Media and Sport; Chris Philp MP, Parliamentary Under-Secretary of State (Minister for Tech and the Digital Economy), Department for Digital, Culture, Media and Sport; Rt Hon Damian Hinds MP, Minister of State (Minister for Security and Borders), Home Office; Sarah Connolly, Director, Security and Online Harms, Department for Digital, Culture, Media and Sport".

Index

References to figures appear in *italic* type.

O

P

R